IAENG TRANSACTIONS ON ELECTRICAL ENGINEERING
VOLUME 1

Special Issue of the International
MultiConference of Engineers and
Computer Scientists 2012

IAENG TRANSACTIONS ON ELECTRICAL ENGINEERING
VOLUME 1

Special Issue of the International
MultiConference of Engineers and
Computer Scientists 2012

Editors

Sio-Iong Ao
International Association of Engineers, Hong Kong

Alan Hoi-shou Chan
City University of Hong Kong, Hong Kong

Hideki Katagiri
Hiroshima University, Japan

Li Xu
Zhejiang University, China

 **World Scientific**

NEW JERSEY · LONDON · SINGAPORE · BEIJING · SHANGHAI · HONG KONG · TAIPEI · CHENNAI

Published by

World Scientific Publishing Co. Pte. Ltd.

5 Toh Tuck Link, Singapore 596224

USA office: 27 Warren Street, Suite 401-402, Hackensack, NJ 07601

UK office: 57 Shelton Street, Covent Garden, London WC2H 9HE

British Library Cataloguing-in-Publication Data
A catalogue record for this book is available from the British Library.

IAENG TRANSACTIONS ON ELECTRICAL ENGINEERING VOLUME 1
Special Issue of the International MultiConference of Engineers and
Computer Scientists 2012

ISBN 978-981-4439-07-7

Printed in Singapore.

PREFACE

A large international conference on Advances in Electrical Engineering was held in Hong Kong, March 14-16, 2012, under the International MultiConference of Engineers and Computer Scientists (IMECS 2012). The IMECS 2012 is organized by the International Association of Engineers (IAENG). IAENG is a non-profit international association for the engineers and the computer scientists, which was founded originally in 1968 and has been undergoing rapid expansions in recent few years. The IMECS conferences serve as good platforms for the engineering community to meet with each other and to exchange ideas. The conferences have also struck a balance between theoretical and application development. The conference committees have been formed with over three hundred committee members who are mainly research center heads, faculty deans, department heads, professors, and research scientists from over 30 countries (http://www.iaeng.org/IMECS2012/committee.html). The conferences are truly international meetings with a high level of participation from many countries. The response that we have received for the congress is excellent. There have been more than eight hundred manuscript submissions for the IMECS 2012. All submitted papers have gone through the peer review process and the overall acceptance rate is 54.6%.

This volume contains 23 revised and extended research articles written by prominent researchers participating in the conference. Topics covered include electrical engineering, circuits, artificial intelligence, data mining, imaging engineering, bioinformatics, internet computing, software engineering, and industrial applications. The book offers the state of art of tremendous advances in electrical engineering and also serves as an excellent reference work for researchers and graduate students working with/on electrical engineering.

Sio-Iong Ao
Alan Hoi-shou Chan
Hideki Katagiri
Li Xu

CONTENTS

LOW-NOISE MEASUREMENTS OF SMALL CURRENTS AND VOLTAGES FOR CHARACTERIZATION OF SEMICONDUCTOR NANOSTRUCTURES AT LOW TEMPERATURES

J. JACOB*

*Institute of Applied Physics, University of Hamburg,
Hamburg, 20355, Germany
*E-mail: jjacob@physnet.uni-hamburg.de

B. FIEDLER[†]

[†]Department of Information and Electrical Engineering, Hamburg University of
Applied Science,
Hamburg, 20099, Germany
[†]E-mail: boris.fiedler@t-online.de

Measuring small currents and voltages with high precision in the presence of noise is challenging. However, many applications in modern solid state physics require smallest excitations in the range of nanovolt and picoampere to investigate intriguing quantum effects in nano structures. We developed a series of pre-amplification and signal-conditioning systems to allow high quality measurements of such signals with emphasis on multi-channel applications as well as signal clarity.

Keywords: Pre-amplifier, signal conditioning, low-noise signal detection.

1. Introduction

Miniaturization of integrated circuits has reached a stage where quantum effects become important. While these effects make it challenging to operate traditional CMOS devices as expected, they also offer interesting new possibilities, for example the joint utilization of the carrier's spin and charge in so-called spintronic devices. However, investigations of such spin-related effects in semiconductor nanostructures require well defined almost perturbationless environments. Therefore measurements are conducted at millikelvin temperatures and only small currents in the nano- and picoampere range and voltages in the micro- and nanovolt range are used. While frequency-selective ac measurements in lock-in technique can significantly increase the dynamic range and detect signals buried in the noise floor the accurate detection is still challenging and sophisticated pre-amplification and signal conditioning is necessary before feeding the data to analog-digital converters in the acquisition system. In most conventional lock-in amplifiers and in all PC-based solutions the signal digitized immediately after a first amplification and a digital-signal processing chip (DSP) is used to compute the lock-in signal (e.g.[1]). While box instruments are a handy solution for one- or two-channel applications, PC-based solutions become interesting in multi-channel applications where rack-space and also costs are important. For example the LabVIEW Multichannel Lock-In Toolkit[2] provides a convenient solution and systems of up to 48 parallel lock-in have been successfully implemented on an eight-core PC.[3] For high-frequency applications the algorithm is also suitable for implementation in an FPGA.[4]

While this approach includes two main components of box-type lock-in amplifiers – A/D conversion and signal processing – and it makes a third component – communication with the computer – obsolete as it is already integrated in the PC, the pre-amplification and conditioning of the signal is missing. Even traditional box-equipment might need more sophisticated pre-amplification and signal conditioning to detect smallest signals. Commercial devices have the disadvantage of being bulky rack equipment while it is desirable to keep signal lines between the experiment and the amplification and conditioning system as short as possible. The need of a case, a power supply, displays and keypads, and a computer interface further increase the price of conventional out-of-stock equipment while modular solutions could significantly improve the cost efficiency and performance.

Therefore we developed cost efficient, modular, and compact pre-amplification and signal conditioning systems to improve the signal quality in challenging measurement scenarios of small currents and voltages.

2. Multi-Channel Pre-Amplification and Signal Conditioning System

We have designed a preamplifier and signal conditioning module capable of serving various applications in our experiments[5] and summarize its functionality here. A schematic of the analog circuit is shown in Fig. 1. Figure 2 shows the digital control circuitry of the module. The module accepts ac- or dc-couple differential and single-ended voltages in a frequency range from dc up to ac signals of 8 kHz and also provides current-to-voltage conversion in this range. Flexible adoption to other ranges is achieved by replacing a small set of capacitors. Grid-frequency influences can be removed by notch

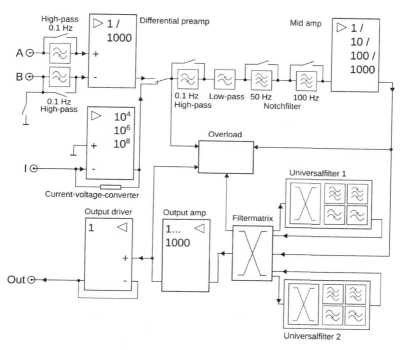

Fig. 1. Block diagram of the analog circuit of the amplifier module. The input signal is connected to port A, B, and/or I. Then the signal is fed either through the differential preamplifier or the current to voltage converter. After that ac or dc coupling is selected, out-of-band noise as well as influences from the grid reduced and second amplification conducted. The third stage consists of two stage-variable filters that are interconnected by a matrix and allow formation of a variety of tunable filter configurations. Right in front of the output driver a continuously variable gain amplification stage allows to match the output level precisely to the input range of the data-acquisition system.

filters at 50(60) Hz and 100(120) Hz. Out-of-band noise above 20 kHz is removed by a low-pass filter. A matrix of two state-variable filters creates high-, low-, and band-pass as well as notch-filters. The maximum combined amplification of the system's amplification stages allows a 10 V output for signals as small as 100 nV at the input. The current-to-voltage conversion allows full 10 V signals at the output for input signals as small as 10 pA. The amplification can be adjusted by switchable gain factors and a variable amplification in the state-variable filters to perfectly match any input signal to the range of the following data-acquisition system.

The modular system houses several pre-amplifier and signal conditioning modules in a mainframe that is also equipped with a controller module for

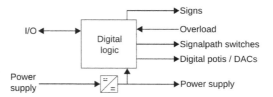

Fig. 2. Block diagram of the control and power supply regulation board. The supply lines from the power supply are received from the backplane connector and are conditioned and converted to the voltages needed on the analog board. A CPLD communicates with the controller module via a SPI protocol over the backplane and controls the switches, potentiometers, and digital-analog converters on the analog board via different protocols.

manual control of the amplifiers as well as computer control, a power supply module, and the backplane interconnecting all subsystems.

2.1. *Input Amplifier Module*

The input amplifier module can amplify differential and single-ended voltages with two user-selectable gain factors. Current-to-voltage conversion with three selectable proportional factors is also available.

Differential Amplifier The differential amplifier is a low noise LT1167 from Linear Technologies[6] with high common mode rejection, small dc offset and a maximum input range of ± 10 V. This stage's circuit can be found in Fig. 3. The gain can be switched from 1 to 1000 by inserting an appropriate 49.45 Ω resistor with a relay into the amplifier's feedback path. For single-ended measurements one input can be connected to ground. The inputs can be floating or referenced to ground and 0.1 Hz high-pass filters can be inserted for ac coupling. A set of JFET transistors in the input leads provide electrostatic discharge and over voltage protection. Automated offset correction via an inverting integrator in the feedback loop or a potentiometer can trim the amplifier's input.

Current To Voltage Converter The current-to-voltage converter uses a low-noise AD 549[7] operational amplifier with very low input bias current. Proportional factors of 10^4 V/A and 10^6 V/A are selected by inserting a 10 kΩ resistance in parallel to a 1 MΩ resistor into the feedback loop as shown in Fig. 4. A proportional factor of 10^8 VA can be achieved by inserting a 1:100 voltage divider at the end of the feedback loop. Diodes provide overload and discharge protection at the input. A voltage follower and an

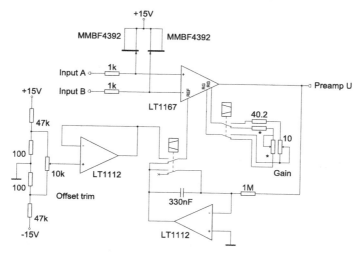

Fig. 3. Differential amplifier in the input stage for ac and dc coupled measurements of differential and single-ended voltage signals

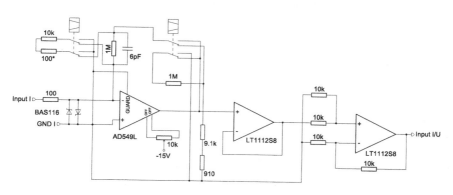

Fig. 4. Current to voltage converter circuit in the input stage.

instrumentation amplifier allow floating measurements. As large resistors at the input would cause large bias currents the ac coupling option has been moved to the following filter section after the amplification.

2.2. Preliminary Filtering

In the preliminary filtering a 0.16 Hz high-pass filter can be inserted for ac coupling in the current-input mode. A 20 kHz fourth-order low-pass filter rejects out-of-band noise without attenuating signals in the measurement

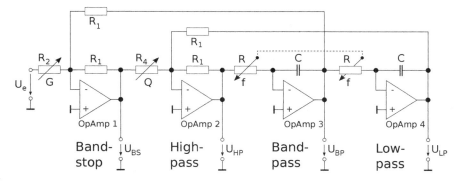

Fig. 5.   Basic circuit of a state variable filter[8]

range up to 8 kHz. Two notch filters at 50(60) Hz and 100(120) Hz remove influences from the grid by subtracting the signal of a very sharp band-pass from the original signal providing strong suppression and minimal influence on the phase. Additional amplification is provided here by a non-inverting amplifier with selectable gain factors of 1, 10, 100, and 1000 by inserting different resistors into the amplifier's feedback loop.

2.3. *Universal Filter Matrix*

The universal filter matrix is the unique feature of this system not found in traditional box-equipment. After the preliminary filtering the signal is fed to two state-variable filters interconnected by a matrix allowing a variety of filter configurations.

State Variable Filter The schematic of a state variable filter based on[8] is shown in Fig. 5. Replacing the analog potentiometer R2 with an AD5293 10-bit digital potentiometer[9] allows digital controlled amplification from 0.5 to 1.5 with 0.01 resolution to match the input range of the data-acquisition system. To allow fine tuning of the Q-factors between 0.4 and 400 as well as cut-off / resonance frequencies from dc to 8120 Hz with 125 mHz resolution the potentiometers R and R4 have been replaced by AD5543 16-bit digital-analog converters.[10] To change the frequency range for different application just two capacitors have to be changed. The implemented circuit is shown in in Fig. 6 The D/A-converters output a current proportional to their reference input, that is in our case hooked to the input signal, and the digitally set resistance of their internal resistor network. By converting the current back into a voltage by an external low-noise and small offset

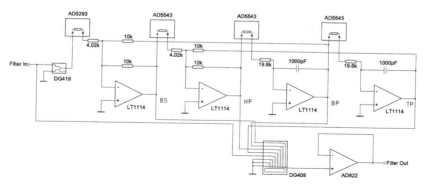

Fig. 6. State variable filter with digitally controllable amplification, Q-factor, and frequency. The additional inverter after the output selector matrix is omitted in the second filter as it is not needed there.

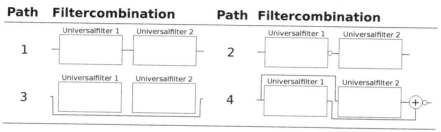

Fig. 7. Overview of the possible combinations of the two state-variable filters by the interconnection matrix.

current-to-voltage converter the circuit matches the behavior of the analog potentiometer but with digital control. The four outputs of the filter provide a high-pass, a low-pass, a band-pass, and a band-stop filtered signal simultaneously to multiplex allowing to select any of theses four or the original signal as the filter's output. The first state-variable filter also provides an inverted output allowing more flexible combinations with the second stage.

Matrix The two state-variable filters are connected to a dual 1:4 analog multiplexer that allows different filter combinations as shown in Fig. 7. As can be seen in Fig. 8 this section includes a high-pass filter to remove any dc offset created in the state variable filters as a correction inside the filter is hard to achieve as the offset changes with the settings of the DA-converters. For dc measurements there is a digital potentiometer in the final impedance converter of the system to tune the offset corresponding to the settings of the state-variable filters according to a calibration table.

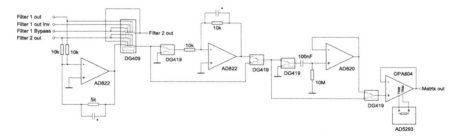

Fig. 8. Matrix circuit connecting the two state-variable filters by a dual 1:4 multiplexer and providing additional offset adjustment.

The set of the two filters connected via the matrix allows for example a narrow fourth-order band-pass filter by cascading both filters with the same settings. Setting one filter to a high- and the other to a low-pass one can create a wide band-pass for measurements requiring a certain bandwidth and at the same time a decent noise reduction. Summing the signals of the two filters set to band-pass configuration but with different frequencies allows detection of a base frequency and its harmonics.

2.4. *Output Stage*

Before the signal is passed to the data acquisition system via a voltage follower with an output driver there is a final amplification stage that provides two gain ranges between 2-102 or 11-1011 via digital potentiometer and a switch for the ranges. The output impedance is 50 Ω and the maximum current that the output can provide is 250 mA.

2.5. *Control and Power Supply*

The pre-amplifier and filter modules is controlled by a CPLD that also is responsible for the communication with the manual and computer control module via the backplane. The backplane connector also provides power to the module in different voltages that are then further conditioned on the module itself.

Control The CPLD provides 0 V/3.3 V digital output to toggle the analog switches in the different stages of the module. For the bi-stable relays mono flops connected to the CPLD trigger driver-ICs to create powerful 5 V pulses. To control the DA-converters and digital potentiometers SPI protocol commands generated in the controller module are passed on to the devices by the CPLD. The chip receives information from the overload control circuits that are connected to pull-down resistors to stay high until pulled down to the low state in case of an overload. Over the backplane the CPLD is connected to the controller module via a SPI-bus and an individual chip-select line that enables distinctive communication with a single amplifier module allowing multiple amplifiers to be connected to the same controller module via the backplane of a 19" rack housing. The controller module features a LCD-Display and a turn-and-push button for user input as well as a USB-connection to the host computer for automated control by a set of procedures of the LabVIEW-based driver.

Power Supply The backplane provides power supply lines for ± 18 V and $+8$ V. The voltages needed with in the amplifier module are various: The operational amplifiers are supplied with ± 15 V; for the relays $+5$ V are necessary; and the DA-converters as well as the digital potentiometers require $+3.3$ V. The ± 18 V are fed through voltage regulators to provide the ± 15 V voltages. As the differential amplifier and the current-to-voltage converter in the input stage need higher accuracy a separate, stabilized and filtered ± 15 V supply is provided. Further cascaded regulators provide the $+5$ V and the $+3.3$ V supply from the $+8$ V input.

2.6. *Circuit Board Design*

The module is split into the analog circuit on one printed circuit board and the digital controller together with the power supply on a second board for farthest possible separation of analog and digital signals to reduce noise. The fact that within the chosen protocol digital signals are only present while a settings change in conducted further reduces digital noise. Within the analog board the digital signals are all routed on one side while the analog signals are on the other side of the double layer board. Similarly on the digital board the power supply is on one side and the digital signals on the other. All free areas on both boards are filled with guard plates. Especially sophisticated guard shielding around the current-to-voltage converter as been place according to Ref. 7.

2.7. *Performance*

The chosen extreme low-noise components result in a calculated input noise of 9.6 nV/$\sqrt{\text{Hz}}$ at a bandwidth of 12 kHz when setting the differential amplifier to a gain of 1000. The corresponding output noise of the amplifier is

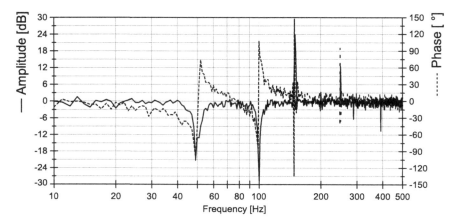

Fig. 9. Frequency response of the preamplifier with the two notch filters inserted into the signal path for a single-ended input voltage of 100 nV and a total gain of 10^5.

1.1 mV and therefore more than a factor of two better than in the previously used commercial lock-in amplifier.[1] The current input noise of 134 fA/$\sqrt{\text{Hz}}$ at 1 kHz for a proportional gain of 10^6 V/A is comparable to the commercial device.[1] The frequency response in Fig. 9 is almost linear in the low-frequency regime with two deep and sharp notch filters on the grid frequency and its first harmonic. In Fig. 10 we show exemplarily the frequency response of the module with one of the universal state-variable filters configured as a band pass with a center frequency of 1 kHz and a Q-factor of 25. The overall gain in this configuration is 10^7 corresponding to an output voltage of 1 V for an input voltage of 100 nV.

3. Battery-Powered Current and Voltage Amplification for Ultra-Low-Noise Measurements

While the just presented pre-amplification and signal conditioning module is suitable for a huge variety of experiments, it is in some cases necessary to further reduce the noise levels and actually even removing noise on the signal line to sample to avoid undesired responses of the system under investigation to such a noisy input signal. This would suggest to implement the pre-amplification and the conditioning of applied signals as close as possible to the sample. In the ideal case one would like to locate the necessary electronics directly at the sample stage inside the low-temperature cryostat. However, not only space is limited here also the extremely low temperatures below 4 K and high magnetic fields of several Tesla would change the behav-

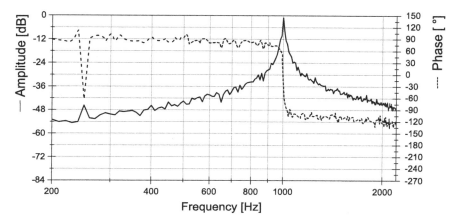

Fig. 10. Frequency response of the preamplifier with one state-variable filter configured as a narrow band pass at 1 kHz with a Q factor of 25.

ior of the employed electronic components. Therefore, in most situations the best compromise is to place the electronics right at the room-temperature connection of the low-temperature setup in a well shielded housing. For such applications we developed a compact, battery-powered pre-amplification system that contains the essential components for a four-point measurement of small currents and voltages to determine the sample's resistance. As shown in the schematic in Fig. 11 the system consists of a differential amplifier for the voltage measurements, a current-to-voltage converter, and voltage dividers. Positioning this system directly at the room-temperature outlet of the low-temperature setup ensures that only large amplified signals are traveling from the system to the data-acquisition devices and that also only large signals have to be send from the data-acquisition system to the measurement setup as they will be transform the the appropriate small but noise sensitive level in the electronics right when entering the low-temperature setup. We have strictly separated the signal ground from the chassis ground in the pre-amplification system to avoid noise contribution from ground loops. The differential amplifier section shown in Fig. 12 is similar to the one in the previously described system and allows different gain settings of 1, 10, 100, and 1000 via inserting different resistors in its feedback loop by setting a jumper accordingly. While commercial lock-in amplifiers yield an input impedance of $10 - 100$ MΩ the very high input impedance of our system of 100 GΩ makes it highly suitable for measurements, where the sample's resistance is varied by a gate leading ultimately

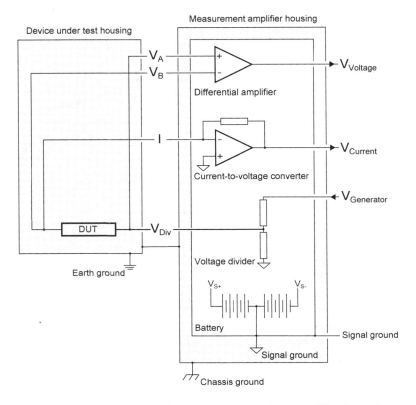

Fig. 11. Block schematic of the battery-powered pre-amplification unit.

to complete pinch-off of the sample. Also for the current-to-voltage converter shown in Fig. 13 we rely on the proven design of the other measurement system and allow in this new application jumper-selectable proportional factors of 10^6 V/A and 10^8 V/A. For the ac voltage applied to the sample a voltage divider with jumper-selectable ratios of 1:1.000, 1:10.000. and 1:100.000 allows large signals of the generator's maximum amplitude to reach the electronics box and transforms them to nanovolt scale for the excitation of the sample. All lines connecting the electronics to the data-acquisition system are fitted with π-filters to reject high-frequency noise. DC lines from the data-acquisition system to the electronics box to provide gate voltages are equipped with additional filters to remove noise from these critical signals. All our low-temperature setups are configured to provide 24 electrical connections between the sample stage and room temperature

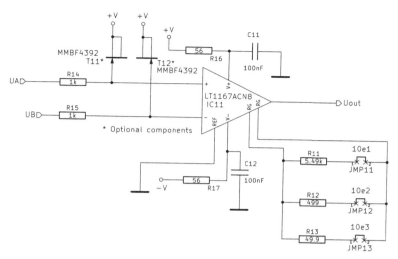

Fig. 12. Circuit layout of the differential amplifier section in the battery-power pre-amplification module.

via a 24-pin Fischer plug. The electronics box is connected via such a plug to the low-temperature setup and allows patching of all its amplifiers and voltage dividers to any of the 24 pins giving access to multiple samples in the low-temperature setup without warming up the system and changing the wiring on the low-temperature side. To eliminate noise from the power supply lines we developed a battery module powering the active electronics components. Two sets of four NiMH eneloop cells provide the positive and negative supply voltages of ± 4.8 V. The voltages are constantly monitored and system health is displayed the the user via a set of LEDs indicating a normal supply voltage of > 4.5 V (green), almost depleted batteries providing $4.0 - 4.5$ V (yellow), and a critical state of < 4.0 V (red) followed by a shut down of the power supply. Changes of the battery voltage within the operation range from 4.8 V to 4.5 V do not show any influence on the measurements. With the chosen batteries and operation limits continuous operation of the pre-amplification system of more then two weeks can be achieved.

4. Conclusion

We developed a universal pre-amplifier and signal conditioning module for a variety of multi-purpose applications that require the detection of small currents and voltages as it can be switched between single-ended and dif-

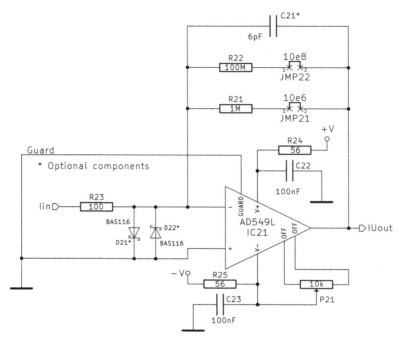

Fig. 13. Circuit of the current-to-voltage converter in the battery-power pre-amplification unit.

ferential voltage signals as well as a current input. Its versatile signal conditioning featuring a matrix-combinable set of two state-variable filters allows easy adaption to many different measurement situations and the ability to combine multiple modules in a central chassis makes the module highly suitable for multi-channel applications.

While this module already exhibits superior noise figures compared with commercial equipment we further improved the signal quality in our battery-powered extremely low-noise preamplifier module. Due to its placement directly at the interface of the low-temperature equipment space is limited and some of the flexibility of the signal-conditioning had to be removed. The signal-to-noise ration and the overall signal quality – and that is the main focus of the module – on the other hand were again significantly improved allowing much finer resolution measurements of even smaller signals. The combination of both modules represents the ideal package for all kinds of current and voltage measurements at low-temperatures on nano structures where very small signal have to be reliably detected.

Acknowledgment

The authors would like to thank Hans Peter Kölzer, Toru Matsuyama, and Horst Biedermann for fruitful discussions and helpful thoughts during the development of the presented circuits. Financial support by the Deutsche Forschungsgemeinschaft via the Graduiertenkolleg 1286 "Functional Metal-Semiconductor Hybrid Devices" and the Project DFG Me916/11-1 "InAs Spin-Filter Cascades" is greatfully acknowledged.

References

1. Stanford Research, Dual-Phase Lock-In Amplifier SR830 http://www.thinksrs.com/products/SR810830.htm.
2. National Instruments, Multi Channel Count Lock-In Amplifier with NI-4472 and DAQmx http://zone.ni.com/devzone/cda/epd/p/id/4532.
3. J. Jacob, Low-Temperature and High-Magnetic Field Laboratory for Spin- and Charge Transport in Semiconductor Nanostructures, unpublished.
4. National Instruments, Lock in Amplifier on LabVIEW FPGA https://decibel.ni.com/content/docs/DOC-1762.
5. J. Jacob and B. Fiedler, *Modular Analog Current and Voltage Preamplifier System with State Variable Filters and Computer Control Interface* (Lecture Notes in Engineering and Computer Science: Proceedings of The International MultiConference of Engineers and Computer Scientists 2012, IMECS 2012, 14-16 March, 2012, Hong Kong, 2012), pp. 1006 – 1011.
6. Linear Technologies, LT1167 - Single Resistor Gain Programmable, Precision Instrumentation Amplifier http://www.linear.com/product/LT1167.
7. Analog Devices, AD549: Ultralow Input-Bias Current Operational Amplifier http://www.analog.com/en/all-operational-amplifiers-op-amps/operational-amplifiers-op-amps/ad549/products/product.html.
8. U. Tietze, C. Schenk and E. Gamm, *Halbleiter-Schaltungstechnik*, 12 edn. (Springer, Berlin, 2002).
9. Analog Devices, AD5293: Single Channel, 1024-Position, 1% R-Tol, Digital Potentiometer http://www.analog.com/en/digital-to-analog-converters/digital-potentiometers/ad5293/products/product.html.
10. Analog Devices, AD5543: 16-Bit DAC in SOIC-8 Package http://www.analog.com/en/digital-to-analog-converters/da-converters/ad5543/products/product.html.

AN INTEGRATED APPROACH TO POWER QUALITY PROBLEMS IN MICRO-GRIDS

TSAO-TSUNG MA

Department of Electrical Engineering, CEECS, National United University
No. 1 Lien-Da, Kung-Ching Li, Miao-Li City, 36003, TAIWAN
Email:tonyma@nuu.edu.tw

This paper describes an integrated approach to power quality problems in micro-grids. A digital signal processor (DSP) based flexible control scheme designed for the distributed generator (DG) inverter to perform multiple control functions is investigated. In particular the attention has been focused on the issues of power quality enhancement in micro-grids with DG inverters. The proposed control algorithm is derived from applying the Clark transformation (a-b-c to α-β) and Park transformation (α-β to d-q) to the power system parameters and the related DG control variables. With the proposed P-Q decoupled controllers designed on the d-q synchronously rotating reference frame, the active and reactive currents injected by the DG inverter can be controlled independently. It is important to note that the new control scheme proposed in this paper does not require a conventional phase locked loop (PLL) in its control circuit thus it has relatively fast dynamic response in real-time power flow regulation. The main advantage of the proposed control approach is that multiple control functions are integrated into a single DG inverter without additional sensing devices and hardware equipment. To verify the advantages and effectiveness of the control scheme, comprehensive numerical simulations and experiments on DSP based hardware setup have been carried out and typical results have been presented with brief discussions.

1. Introduction

In recent years, the quantity of distributed generators (DG) connected to power distribution networks has rapidly increased. A number of reasons can explain this trend; i.e., environmental concerns, electricity business restructuring, the fast developments of small scale power generation technologies and micro-grid related devices and systems. In practice, the input energy units of DG systems can be constructed with various renewable and green energy sources; however, the real power output from these energy resources is intrinsically unstable. As a result, the complexity of design and operating issues and the requirement of control means have been greatly increased. With the inevitable trend of increasing renewable energy sources and DG installations, it is a crucial need to develop new control strategies for the secure operation and optimal management of new power grids embedded with DG and micro-grid units. Potential control schemes aim to maintain or even to improve the overall system reliability and quality. To achieve the above objectives, it is well known that power electronics technology and advanced digital controllers play an important role [1]-[6]. In the open literature, a huge number of DG application examples can be found. Feasible configurations of standalone PV based power generation systems have been reported. Some typical systems based on the fuel cell devices were investigated in [7]-[9], while other practical DG systems were designed and analyzed in [10], [11]. A typical DG system based on wind power generator (WTG) was presented in [12]. It has been reported that as the penetration level of various DG and grid-tied micro-grid systems are increasing the voltage stability, reactive power compensation and power quality related problems have become vital issues in the control of distribution systems. Reactive current increases the distribution system losses, limits the active power transfer capability, reduces the system power factor, and can even cause large-amplitude variations in the load-side voltage. It should be noted that in some cases fast changes in the reactive power consumption of large loads can cause voltage amplitude oscillations. This might also lead to a change in system real power demand resulting in power oscillations [13]-[15]. Other critical issues challenging power engineers in operation and control of interconnected power systems embedded with multiple DG systems may include real-time optimization control, power quality control, dynamic stability and protection coordination.

Although the main objective of DG in micro-grid systems is to provide active power; nonetheless, by means of power electronic systems and digital controllers with properly designed control schemes, reactive power can be fully compensated and DG inverter systems can also provide additional control functions, such as the compensation of load harmonic currents of active power filters (APF) if the power rating of the DG inverter is allowed. Theoretically, DG systems can be connected to the micro-grid network in series or in shunt; however, because the possible compensated quantities; e.g. reactive power or harmonics, are directly related to the currents, shunt type topology is more realistic in designing practical DG systems as it can effectively injects the required compensating currents at the point of common coupling (PCC). Therefore, the shunt type DG inverter is used in this paper for the theoretical analysis and design of related controllers. In practical applications, the three-phase voltage source inverter (VSI) has been widely used for interfacing between DG and micro-grid networks. To achieve a multi-functional DG inverter (MDGI), the command signals for the VSI, which are current signals in nature, may include the information of active power supplied from renewable energy sources and reactive power required to compensate the voltage fluctuation at load-side as well as harmonic currents. In this paper, the hysteresis current control algorithm is used for its fast dynamic response, accurate performance and ease of implementation. It is important to note that to realize a simple, cost-effective and high-performance MDGI system the phase locked loop (PLL) is not used in the proposed control circuit. The performance of the proposed control scheme is firstly simulated using PSIM software and followed by a set of tests on DSP based hardware implementations. The results are then presented and briefly discussed to demonstrate the feasibility and effectiveness of the proposed control scheme.

2. Micro-grid and Control Techniques

2.1. *Micro-grid Concepts and Power Quality Issues*

A conceptual micro-grid architecture is shown in Figure 1. It consists of a group of radial feeders, which could be part of a distribution system or the electrical power system of residential buildings. Of the control requirements in operating micro-grids, which are normally issued by the grid operators, power quality has recently gained a lot of attention due to excessive non-linear and unbalanced loads used in such power systems. Without proper measures, these power quality problems may easily cause system failures. This is because the nonlinear or unbalanced loads can represent a very high percentage of the total load in small-scale power systems like a micro-grid. Thus, the problem related to power quality has become a particular concern in operation and control of micro-grids.

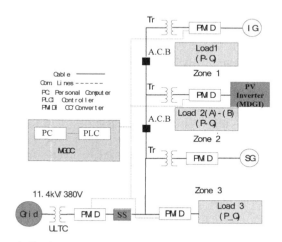

Figure 1. Simple system diagram of a micro-grid with various DG inverters.

As can be seen in Figure 1, there is a static switch (SS) installed at the point of connection to the utility grid which is controlled to separate the micro-grid from a faulty utility grid in a fraction of a cycle. In a micro-grid, some feeders may form a number of distinct zones with various sensitive loads, e.g., Load1, Load2 (A)-(B) as shown in Figure 1 which may require local DG on renewable energy sources (RES) and certain energy storage systems. The non-critical load feeder (Load3) is normally not equipped with any local generation. In this paper, the PV Inverter shown in Figure 1 is used to demonstrate the proposed MDGI concept and Zone1 to Zone3 can be islanded from the grid using the SS if so desired.

2.2. *Control Techniques*

As mentioned in the introduction, the current control techniques of VSI presented in this paper is based on analysis of voltage and current vector components in a special synchronous d-q-reference frame. To decompose voltage and current components in a rotating reference frame, calculation of instantaneous angle of voltage or current is needed. Various phase-angle detecting methods have already been developed and reported. To obtain this angle, a PLL is commonly used in voltage source inverter's control loop [16]-[20]. It should be noted that using PLL has some disadvantages; such as problems due to synchronization of DG with the grid and elimination of a wide range of frequencies which is not favorable in DG applications, especially when the function of active power filter is activated. In addition, PLL is very sensitive to noises and disturbances. In the proposed control algorithm, instantaneous angle of load voltage is calculated directly by decomposing voltage vector components in a rotating reference frame. Removing PLL from control circuit of voltage source inverter introduces a simple and reliable control method for DG systems. In this paper, with the proposed d-q method the synchronization problem can be resolved and a better dynamic response of DG inverter can be achieved.

2.3. *Calculation of Reference Current Commands*

The derivation of mathematical models of the nonlinear shunt DG inverter system is based on the equivalent circuit shown in Figure 2. To achieve a reliable control, voltage and current components are firstly obtained in a stationary reference frame and related mathematical manipulations are presented as follows:

$$v_a - L\frac{di_a}{dt} - i_a R - e_a = 0 \tag{1}$$

$$v_b - L\frac{di_b}{dt} - i_b R - e_b = 0 \tag{2}$$

$$v_c - L\frac{di_c}{dt} - i_c R - e_c = 0 \tag{3}$$

Assuming a balanced three-phase system, one can define the following two matrices:

$$P = \begin{bmatrix} \cos\theta & \cos(\theta-120°) & \cos(\theta+120°) \\ -\sin\theta & -\sin(\theta-120°) & -\sin(\theta+120°) \end{bmatrix} \tag{4}$$

$$p^{-1} = \frac{2}{3}\begin{bmatrix} \cos\theta & -\sin\theta \\ \cos(\theta-120°) & -\sin(\theta-120°) \\ \cos(\theta+120°) & -\sin(\theta+120°) \end{bmatrix} \tag{5}$$

Then, the transformations between a-b-c and d-q frames can be performed using the following equations.

$$v_{dq} = pv_{abc} \tag{6}$$

$$i_{dq} = pi_{abc} \tag{7}$$

$$v_{abc} = p^{-1}v_{dq} \tag{8}$$

$$i_{abc} = p^{-1}i_{dq} \tag{9}$$

Using (1)-(7), the following d-q model can be easily derived.

$$e_d = -(Ri_d + L\frac{di_d}{dt}) + \omega L i_q + v_d \tag{10}$$

$$e_q = -(Ri_q + L\frac{di_q}{dt}) - \omega L i_d \tag{11}$$

$$\begin{bmatrix} v_d \\ 0 \end{bmatrix} - L\frac{d}{dt}\begin{bmatrix} i_d \\ i_q \end{bmatrix} - \omega L\begin{bmatrix} -i_q \\ i_d \end{bmatrix} - R\begin{bmatrix} i_d \\ i_q \end{bmatrix} - \begin{bmatrix} e_d \\ e_q \end{bmatrix} = 0 \tag{12}$$

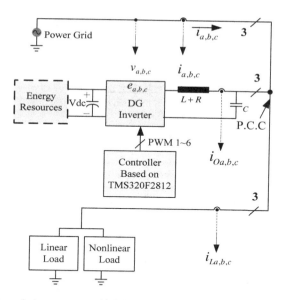

Figure 2. A test system with DG inverter systems and control signals.

It is clear that to calculate the injected current components of DG system to the grid, the related current and voltages signals must be transformed to synchronously rotating reference frame, i.e. in d-q components. In this transformation d-axis vector is normally assumed in the same direction as the vector of grid voltage. With this consideration, vertical component of voltage (d-component) in rotating synchronous reference frame is always zero. Figure 3 and 4 respectively show the voltage and current components in stationary and rotating synchronous reference frames. Transformation matrices based on Park and Clark equations are given in (4)-(9).

According to Figure 4, the q-component of voltage in stationary and rotating synchronous reference frame can be calculated as:

$$v_q = \left| \vec{v}_{dq} \right| = \left| \vec{v}_{\alpha\beta} \right| = \sqrt{v_\alpha^2 + v_\beta^2} \qquad (13)$$

$$\theta = \tan^{-1} \frac{v_\beta}{v_\alpha} \qquad (14)$$

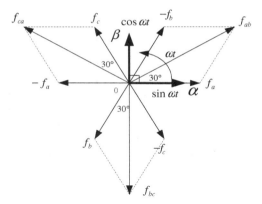

Figure 3. The voltage and current components in stationary reference frames.

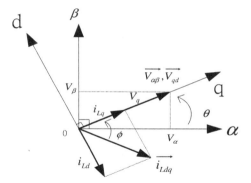

Figure 4. The voltage and current components in rotating synchronous reference frames.

It follows that one can use the above equations to easily obtain the mathematical formulations of real and reactive power in d-q frame.

$$\begin{bmatrix} P_L \\ Q_L \end{bmatrix} = \begin{bmatrix} V_\alpha i_{l\alpha} + V_\beta i_{l\beta} \\ -V_\beta i_{l\alpha} + V_\alpha i_{l\beta} \end{bmatrix} = \begin{bmatrix} V_q i_{lq} + V_d i_{ld} \\ V_q i_{ld} - V_d i_{lq} \end{bmatrix} \qquad (15)$$

$$\begin{bmatrix} P_L \\ Q_L \end{bmatrix} = \begin{bmatrix} V_q i_{lq} \\ V_q i_{ld} \end{bmatrix} \qquad (16)$$

Calculation of the load currents in d-q reference frame by (7) makes it possible to separate fundamental and harmonic components of the load currents if there is any. Using a proper low-pass filter, the current commands can be separated into DC and alternative components expressed as follows:

$$\begin{bmatrix} i_{ld}^* \\ i_{lq}^* \end{bmatrix} = \begin{bmatrix} i_{ld,DC}^* + i_{ld,h} \\ i_{lq,DC}^* + i_{lq,h} \end{bmatrix} \qquad (17)$$

Based on the proposed control algorithm, the overall current control signals can be constructed in d-q synchronous reference frame as expressed in (18).

$$\begin{bmatrix} i_{cd}^* \\ i_{cq}^* \end{bmatrix} = \begin{bmatrix} i_{ld,DC(DG)}^* + i_{ld,h} \\ i_{lq,DC(DG)}^* + i_{lq,h} \end{bmatrix} \qquad (18)$$

In (18), the i_{cd}^* and i_{cq}^* are d-q current commands for the DG inverter. $i_{ld,DC(DG)}^*$ and $i_{lq,DC(DG)}^*$ are used to control respectively the desired output real and reactive power of the DG inverter, while $i_{ld,h}$ and $i_{lq,h}$ are respectively the harmonic components of d-axis and q-axis load currents to be compensated if the APF function is activated. The α and β components of the d-q current commands can then be obtained as:

$$\begin{bmatrix} i_{c\alpha}^* \\ i_{c\beta}^* \end{bmatrix} = \frac{1}{V_d} \begin{bmatrix} V_\alpha & -V_\beta \\ V_\beta & V_\alpha \end{bmatrix} \begin{bmatrix} i_{cd}^* \\ i_{cq}^* \end{bmatrix} \qquad (19)$$

Finally, the overall three-phase current commands for the DG inverter to perform the desired control functions can be reached as follows.

$$i_c^* = \begin{bmatrix} i_{ca}^* \\ i_{cb}^* \\ i_{cc}^* \end{bmatrix} = \mathbf{T}^{-1} \begin{bmatrix} i_{c\alpha}^* \\ i_{c\beta}^* \end{bmatrix} \qquad (20)$$

In the open literature, many current control methods for three phase systems have been proposed. Among them, a current control scheme using hysteresis regulator is used in this paper. The typical advantages of hysteresis current control are its simplicity in implementation and the fast dynamic response of its current loop.

3. Case Studies and Results

To investigate the detailed dynamics of the DG inverter system and to validate the proposed DSP based integrated control scheme, a set of simulation studies based on a simple distribution network connected with a nonlinear DG inverter system as shown in Figure 5 is firstly carried out in PSIM environment. Figure 5 (a) to (b) show the possible routes of the controlled power flow. In this study, it is considered that for the whole period of simulations the local loads are fed by both the main source of the power grid and the DG. During simulation process active power which is delivered from DG link is considered in three cases, i.e., zero output, with the output just the same as that of the local loads and with the output higher than that of the local load demand.

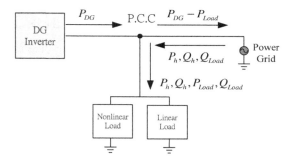

Figure 5 (a). The function of real power generation in a DG.

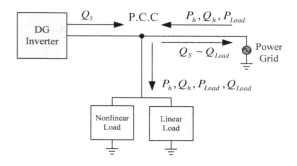

Figure 5 (b). The function of reactive power generation in a DG.

The above arrangement makes it possible to evaluate the capability of DG link to track the fast change in the real and reactive power required by the load independently. To simulate a specific operation scenario, a fixed load type and real power demand are assumed and the harmonic distortion of current waveform are calculated and compared in various control conditions. Since the principle of proposed current control technique is based on separating active and reactive current components in rotating synchronous reference frame known as the d-q components, in all conditions only phase-a parameters (including voltage and current) are shown. To demonstrate the performance of the proposed DG inverter in compensating total load harmonic currents, source current, load current and the output current of the DG are shown simultaneously. The following part of the paper presents the details of simulation cases for various level of output real power from the DG.

3.1. Simulation Cases and Results

In this simulation case, the DG link is connected to the network at t=0.0 sec. At this moment a full-wave AC/DC converter with the output of 156V/500W is added to PCC and it is removed at t=0.6 sec. Figure 6, Figure 7 and Figure 8 respectively show the waveforms of the related voltages and currents in various DG operating conditions. It can be clearly seen in Figure 6 that after the connection of DG the source current becomes sinusoidal and the harmonic currents are fully provided by the DG link. In Figure 7, the nonlinear load demand is fully provided by the DG with an output real power set to 500W. As shown in Figure 8, when the output power of the DG is increased to 2000W there is a real power about 1500W feeding backward to the power grid.

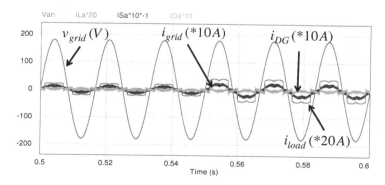

Figure 6. The grid phase-a voltage, current waveforms of load, source and the harmonic currents provided by the DG inverter (P_DG=0W).

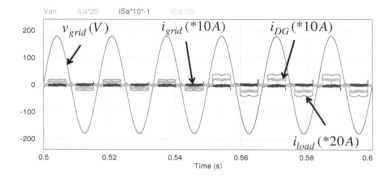

Figure 7. The grid phase-a voltage, current waveforms of load, source and the harmonic currents provided by the DG inverter (P_DG=500W).

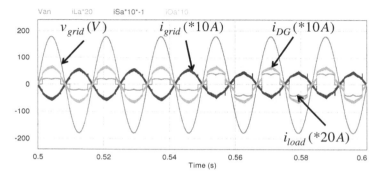

Figure 8. The grid phase-a voltage, current waveforms of load, source and the harmonic currents provided by the DG inverter (P_DG=2000W).

3.2. Experimental Tests and Results

In this study, the proposed control scheme is experimentally tested as configured in Figure 9. In the hardware setup, a digital controller based three phase grid-connected inverter and a set of nonlinear load module are used. Test conditions and parameters are arranged exactly the same as that used in simulation studies. All controllers proposed in this paper are implemented with TI DSP2812. The sensed currents and voltages acquired to the DSP and the control signals output to the driving circuit are using home-made signal acquisition circuits. The arrangement of control signals are shown in Figure 10. Both of the sampling frequency and the switching frequency are set at 24 kHz. Figure 11 to Figure 13 show a set of experimental results regarding the measured voltages and currents shown in the previous simulation studies.

Figure 9. The experimental setup of the proposed DSP based DG inverter system.

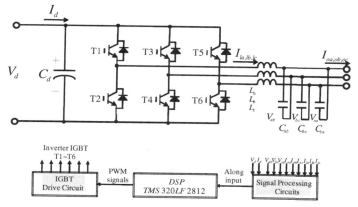

Figure 10. The arrangement of control signals for the DSP based DG inverter system.

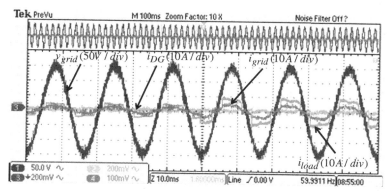

Figure 11. The measured grid phase-a voltage, current waveforms of load, source and the harmonic currents provided by the DSP controlled DG inverter (P_DG=0W).

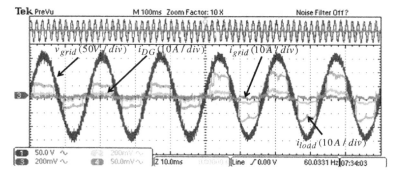

Figure 12. The measured grid phase-a voltage, current waveforms of load, source and the harmonic currents provided by the DSP controlled DG inverter (P_DG=500W).

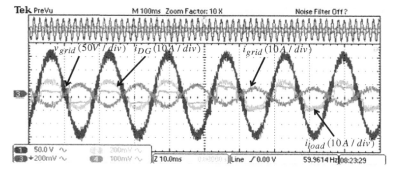

Figure 13. The measured grid phase-a voltage, current waveforms of load, source and the harmonic currents provided by the DSP controlled DG inverter (P_DG=2000W).

4. Conclusion

This paper has presented a novel DSP based integrated control scheme for a conventional DG inverter to perform various control functions without additional sensing devices or hardware equipment. It is important to note that the proposed direct d-q-axis current control method has fast dynamic response in tracking harmonic current variations since the control loops of active and reactive current components are considered independent. Using the proposed DSP based control method, besides the basic real power control function the DG inverter system can also be considered as a new alternative for performing functions of a D-STATCOM and a distributed APF in power distribution networks. Based on the experimental and simulation results obtained the feasibility and effectiveness of the proposed control concept have been fully verified.

Acknowledgments

This work was supported in part by the National Science Council of Taiwan, R.O.C. through: NSC 100-2221-E-239-001.

References

1. J. Arai, K. Iba, T. Funabashi, Y. Nakanishi, K. Koyanagi and R. Yokoyama, *IEEE Circuits and Systems Magazine*, Vo. **8**, No. 3 (2008).
2. O. Alonso, P. Sanchis, E. Gubia and L. Marroyo, *IEEE Power Electronics Specialist Conference*, Vol. **1** (2003).
3. Mudathir F. Akorede, Hashim Hizam, Ishak Aris, Mohd Zainal and A. Ab. Kadir, *IREE*, Vol.**5** , No 2 (2010).
4. P. Bhusal, A. Zahnd, M. Eloholma and L. Halonen, *IREE*, Vol. **2**, No. 1 (2007).
5. H. Laaksonen, A. Mohamed, *IREE*, Vol. **3**, No. 3 (2008).
6. A.A. Salam, G. Barakat, M.A. Hannanand and H. Shareef, *IREE*, Vol. **5**, No. 5 (2010).
7. R. Belfkira, G. Barakat and C. Nichita, *IREE*, Vol. **3**, No.5 (2008).
8. S. Mazumder, R. Burra, R. Huang, M. Tahir, K. Acharya, *IEEE Transactions on Industrial Electronics*, Vol. **1**, No.99 (2010).
9. T. A. Nergaard, J. F. Ferrell, L. G. Leslie and J. S. Lai, *IEEE Power Electronics Specialists Conference*, Vol. **1** (2002).
10. J. Selvaraj, N.A. Rahim, *IEEE Transactions on Industrial Electronics*, Vol.**56**, No.1 (2009).
11. H. Patel, V. Agarwal, *IEEE Transactions on Energy Conversion*, Vol.**24**, No.1 (2009).
12. Y. Jia, Z. Yang and B. Cao, *IEEE Power System Technology Conference*, Vol. **1** (2002).
13. A.J. Watson, P.W. Wheeler, J.C. Clare, *IEEE Transactions on Industrial Electronics*, Vol.**54**, No.6 (2007).
14. L. Gyugyi, *Proceedings of the IEEE*, Vol. **76**, No. 4 (1988).
15. J.G. Singh, S.N. Singh, S.C. Srivastava, *IEEE Transactions on Power Systems*, Vol.**22**, No.4 (2007).
16. D. Yazdani, A. Bakhshai, G. Joos, M. Mojiri, *IEEE Transactions on Power Electronics*, Vol.**23**, No.4 (2008).
17. A. Gebregergis, P. Pillay, IEEE Transactions on Industry Applications, Vol.**46**, No.1 (2010).
18. *Tsao-Tsung Ma*, Power Quality Enhancement in Micro-grids Using Multifunctional DG Inverters, *Lecture Notes in Engineering and Computer Science: Proceedings of The International MultiConference of Engineers and Computer Scientists 2012, IMECS 2012, 14-16 March, 2012, Hong Kong, pp996-1001.*
19. M. Karimi-Ghartemani, IEEE *Trans. Circuits Syst.*, Vol.**53**, No.**8** (2006).
20. C.J. Gajanayake, D.M. Vilathgamuwa, P. C. Loh, R. Teodorescu, F. Blaabjerg, *IEEE Transactions on Energy Conversion*, Vol.**24**, No.3 (2009).

DISCRIMINATING AMONG INRUSH CURRENT, EXTERNAL SHORT CIRCUIT AND INTERNAL WINDING FAULT IN POWER TRANSFORMER USING COEFFICIENT OF DWT[*]

JITTIPHONG KLOMJIT[†]

Department of Electrical Engineering, Faculty of Engineering, King Mongkut's Institute of Technology Ladkrabang, Bangkok, 10520, Thailand, knatthap@gmail.com

ATTHAPOL NGAOPITAKKUL

Department of Electrical Engineering, Faculty of Engineering, King Mongkut's Institute of Technology Ladkrabang, Bangkok, 10520, Thailand, knatthap@gmail.com

This paper proposes a technique for discriminating among inrush current, external fault and internal winding fault of three-phase two-winding transformer which variations of coefficients of high frequency component obtained from DWT of differential current are analyzed. The maximum coefficient details of DWT are performed as comparison indicator. Various cases based on Thailand electricity transmission and distribution systems are studied to verify the validity of the proposed algorithm. Results show that the proposed technique has good accuracy in the considered system

1. Introduction

To guarantee safety and stability of power grid operating, a precise protection scheme is required. Generally, power transformers can be protected by overcurrent relays, pressure relays and differential relays depending on purposes. The differential relaying principle is used for protection of medium and large power transformers. The differential principle, as applied for protecting power transformers, can be described with the help of Figure 1. The levels of currents on the primary and secondary sides of the power transformer are reduced by the current transformers (CTs). The output of these CTs are compared. The ratios of primary and secondary CTs are selected such that each CT produces the same secondary current for nominal line current. Net current through the operating coil of differential relay is zero for normal operation and

---

[*] This work is supported by National Research Council of Thailand and King Mongkut's Institute of Technology Ladkrabang.
[†] Work partially supported by grant of the National Research Council of Thailand and King Mongkut's Institute of Technology Ladkrabang.

external faults. An internal fault in the power transformer upsets this balance and causes a current to flow in the relay's operating coil. This is shown in Figure 1(b).

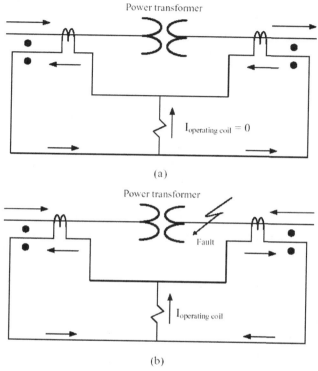

Figure 1. Basic differential scheme for (a) no-fault and (b) internal fault conditions of a power transformer.

However, in some cases, the CTs can be saturated or their ratios are not matched perfectly. This causes the differential signal to become considerably large even though there is no internal fault. In order to prevent false tripping in these cases, in the literature for fault detection, several decision algorithms have been developed to be employed in the protective relay [1-5]. They (most of them) have different solutions and techniques [1-10]. In [1], high frequency component of transformer terminal current technique is proposed for detecting internal faults in power transformers. A digital technique that uses positive- and negative-sequence models of the power system in a fault-detection algorithm is presented in [2]. An application of a finite impulse response ANN (FIRANN) as differential protection for a three-phase power transformer is proposed in [3]. In

[4], this paper describes a new approach for transformer differential protection that ensures security for external faults, inrush, and over-excitation conditions and provides dependability for internal faults. The development of a wavelet–based scheme, for distinguishing between transformer inrush currents and power system fault currents is presented in [5]. A technique for discriminating fault currents from inrush currents is presented in [6]. The performance of the technique is checked from simulation of a 132/11 kV transformer, connected to a 132 kV power system. A new relaying fuzzy logic algorithm to enhance the fault detection sensitivities of conventional techniques is proposed in [7]. The relaying algorithm consists of flux-differential current derivative curve, harmonic restraint, and percentage differential characteristic curve. In [8], a new algorithm based on processing differential current harmonics is proposed for digital differential protection of power transformers. This algorithm has been developed by considering different behaviors of second harmonic components of the differential currents under fault and inrush current conditions. In [9], an Equivalent Instantaneous Inductance (EII)-based scheme is proposed to distinguish the inrush current from internal faults in power transformers, which is derived from the inherent difference of the magnetic permeability, due to the saturation and un-saturation, in the transformer iron core between the inrush current and an internal fault. In [10], this paper describes a new approach for transformer differential protection that ensures security for external faults, inrush and over-excitation conditions and provides dependability for internal faults. As a result, most research works are interested in only the effects from magnetizing inrush current and the discrimination between magnetizing inrush current and internal faults [5-10], and etc.

In addition, the traditional method of signal analysis is based on Fourier transform. However, Fourier transform is not suited to fault transient analysis because the desired information may be not only located in the frequency domain but also in the time domain. Besides, wavelet transform can be used in signal analysis. This mathematical technique is not intended to replace Fourier transform technique in analysing steady state signals, it is just an alternative tool for analysing non-stationary or non-steady state signals.

In previous research works [11], in order to discriminate between external fault and internal fault in power transformer, the comparisons of the coefficients of discrete wavelet transform (DWT) have been performed. The proposed decision algorithm can give more satisfactory results for separation between internal fault and external fault but case studies were verified with power transformer which is connected with wye-wye. In a fact, power transformer which is connected with delta-wye is widely employed more than power transformer which is connected with wye-wye in power system so decision algorithm should prove to be more complicated to analyze.

Therefore, this paper is interested in the decision algorithm for detecting and discriminating among inrush current, internal fault and external fault for power transformer. A decision algorithm is based on wavelet transform as an alternative or improvement to the existing protective relaying functions. The construction of the decision algorithm is detailed and implemented with various case studies based on Thailand electricity transmission and distribution systems.

2. Wavelet Transform

The wavelet transform is a tool that cuts up data or functions or operators into different frequency components, and then studies each component with a resolution matched to its scale. The advantage of the transform is that the band of analysis can be fine adjusted so that high frequency components and low frequency components are detected precisely. Results from the wavelet transform are shown both in time domain and in frequency domain. The wavelet transform, which has a change in the analysis scaled by the factor of two, is called discrete wavelet transform (DWT) as shown in Equation 1.

$$DWT(m,n) = \frac{1}{\sqrt{2^m}} \sum_k f(k) \psi \frac{n - k2^m}{2^m} \tag{1}$$

where,

$\psi \dfrac{n - k2^m}{2^m}$ = mother wavelet

3. Power System Simulation using EMTP

To study internal faults of the transformer, Bastard et al [12] proposed modification of the BCTRAN subroutine. Normally, the BCTRAN uses a matrix of inductances with a size of 6x6 to represent a transformer, but with the internal fault conditions, the matrix is adjusted to be a size of 7x7 for winding to ground faults. However, the effects of high frequency components which may occur during the faults are not included in such a model. In this paper, the combination between the transformer models proposed by Bastard et al [12] and the high frequency model including capacitances of the transformer recommended by IEEE working group [13] is used for simulations of internal faults the transformer windings.

The process for simulating internal faults based on the BCTRAN routine of EMTP can be summarized as follows:

1st step: Compute matrices]R [and]L [with a size of 6x6 to represent a power transformer from manufacture test data [14] without considering the internal faults .

2nd step: Modify matrix of]R [and]L[to be a size of 7x7 for winding to ground faults and of 8x8 for interturn faults.

3rd step: The inter-winding capacitances and earth capacitances of the HV and LV windings can be simulated by adding lumped capacitances connected to the terminals of the transformer.

The scheme under investigations is a part of Thailand electricity transmission and distribution system as depicted in Figure 2. A 50 MVA, 115/23 kV three-phase two-winding transformer was employed in simulations with all parameters and configuration provided by a manufacturer [14].

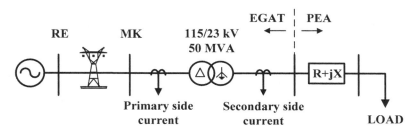

Figure 2. The system used in simulations studies [15].

It can be seen that the transformer which is a step down transformer is connected between two subtransmission sections. To implement the transformer model, simulations were performed with various changes in system parameters as follows:

- The angles on phase A voltage waveform for the instants of fault inception were 0°-330° (each step is 30°).

- For the internal winding faults, the fault positions as shown in Figure 4, were designated on any phases of the transformer windings (both primary and secondary) at the length of 10%, 20%, 30%, 40%, 50%, 60%, 70%, 80% and 90% measured from the line end of the windings.

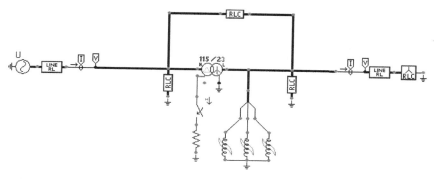

Figure 3. Internal fault model implemented in ATP/EMTP.

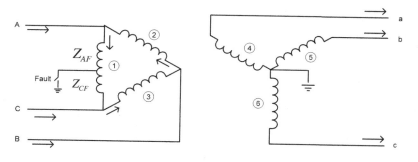

Fault high voltage Coil 1

Figure 4. The modification on ATP/EMTP model for a three-phase transformer with winding to ground faults.

For simulations of external short circuit occurring at the transmission lines at both sides of the transformer, case studies were varied as follows:

- The angles on phase A voltage waveform for the instants of fault inception were 0°-150° (each step is 30°).
- Types of faults were single line to ground, double lines to ground, line to line and three-phase faults (AG, BG, CG, ABG, BCG, CAG, AB, BC, CA, ABC).
- The fault locations on the transmission lines were at the length of 10%, 20%, 30%, 40%, 50%, 60%, 70%, 80% and 90%.
- Fault resistance was 5 Ω.

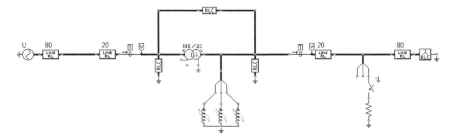

Figure 5. Components of a proposed simulation model in case of external short circuit.

For simulations of inrush current, the transformers were initially energized on high voltage side from no load condition. Case studies were varied as follows:

- The angles on phase A voltage waveform for the instants of fault inception are 0°-150° (each step is 30°).
- The magnetizing flux are 80%, 90%, 100% 110%, and 120%.

4. Fault Detection Algorithm

The primary and secondary current waveforms can be simulated using ATP/EMTP, and these waveforms are interfaced to MATLAB/Simulink for a construction of fault diagnosis process. With fault signals obtained from the simulations, the differential currents, which are a deduction between the primary current and the secondary current in all three phases as well as the zero sequence, are calculated, and the resulted current signals are extracted using the Wavelet transform. The mother wavelet daubechies4 (db4) [16, 17, 18] is employed to decompose high frequency components from signals. From the differential current of all three phases, coefficients from each scale of Wavelet transform are considered and employed in the fault detection with several trial and error processes as shown in Figure 6.

From Figure 7, the coefficients of the signals obtained from the DWT are squared. The comparison of the coefficients from each scale is investigated. The result is clearly seen that when fault occurs, the coefficients of high frequency components have a sudden change compared with those before an occurrence of the faults as illustrated in Figure 7. This sudden change is used as an index for the occurrence of faults.

From Figure 7, it is clearly seen that the coefficients of high frequency components, when fault occurs, have a sudden change compared with those before an occurrence of the faults but the coefficient in each scale of the DWT does not obviously change as shown in Figure 7(a)

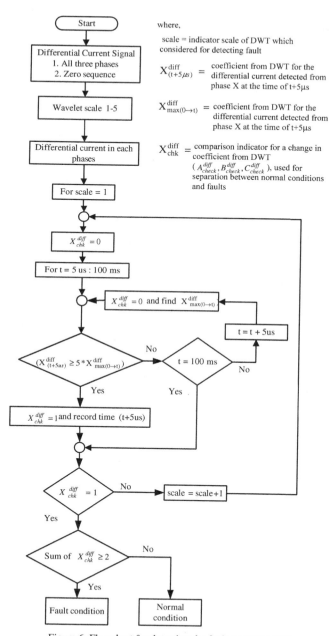

where,

scale = indicator scale of DWT which considered for detecting fault

$X^{\text{diff}}_{(t+5\mu s)}$ = coefficient from DWT for the differential current detected from phase X at the time of t+5μs

$X^{\text{diff}}_{\max(0\to t)}$ = coefficient from DWT for the differential current detected from phase X at the time of t+5μs

$X^{\text{diff}}_{\text{chk}}$ = comparison indicator for a change in coefficient from DWT ($A^{diff}_{check}, B^{diff}_{check}, C^{diff}_{check}$), used for separation between normal conditions and faults

Figure 6. Flowchart for detecting the fault condition [18].

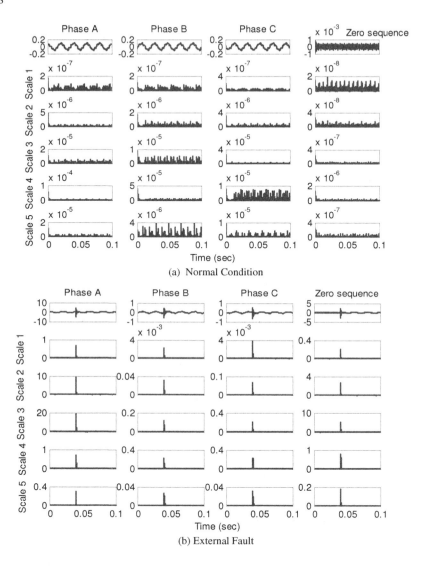

(a) Normal Condition

(b) External Fault

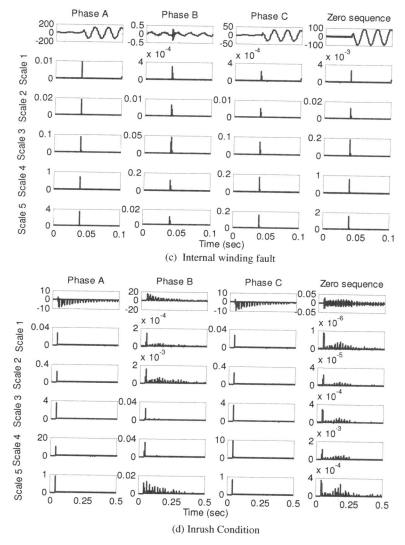

(c) Internal winding fault

(d) Inrush Condition

Figure 7. Wavelet transform of differential currents.

After applying the fault detection algorithm, DWT are applied to the quarter cycle of current waveforms after the fault inception. The coefficients of scale 1 obtained using the DWT are used for discriminating among inrush current, external short circuit and internal winding fault as shown in Figure 8.

40

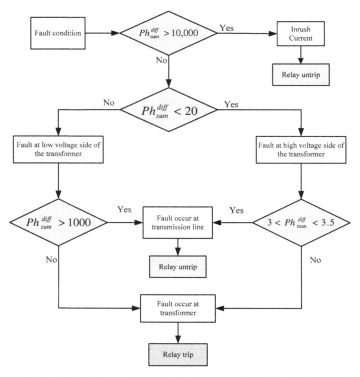

Figure 8. Flowchart for discriminating among inrush current, external fault and internal fault.

where,

$$I_{XZ,\max(post)}^{diff} = \frac{X_{\max(post)}^{diff}}{Z_{\max(post)}^{diff}}$$

$X_{\max(post)}^{diff}$ = maximum value of coefficient from DWT of differential current for phase X at the time of ¼ cycles after detecting faults

$Z_{\max(post)}^{diff}$ = maximum value of coefficient from DWT of zero sequence current at the time of ¼ cycles after detecting faults

$Ph_{\max}^{diff}$ = maximum value of comparison indicators ($I_{AZ,\max(post)}^{diff}, I_{BZ,\max(post)}^{diff}, I_{CZ,\max(post)}^{diff}$) used in classifying the fault condition

Ph_{sum}^{diff} = summation value of comparison indicators ($I_{AZ,\max(post)}^{diff}, I_{BZ,\max(post)}^{diff}, I_{CZ,\max(post)}^{diff}$) used in classifying the fault condition

For classifying the fault condition, the coefficient from scale1 of DWT for differential current and zero sequence current waveforms is used in the decision algorithm. The results are shown as Tables 1, 2 and 3.

Table 1. Results of classifying for external faults.

Differential Current					Result	
$I_{AZ,\max(post)}^{diff}$	$I_{BZ,\max(post)}^{diff}$	$I_{CZ,\max(post)}^{diff}$	Ph_{sum}	$Ph_{\max}$	Condition	Relay
3.33	1.14e-2	1.88e-2	3.36	3.33	External	Un-Trip

Table 2. Results of classifying for internal winding faults.

Differential Current					Result	
$I_{AZ,\max(post)}^{diff}$	$I_{BZ,\max(post)}^{diff}$	$I_{CZ,\max(post)}^{diff}$	Ph_{sum}	$Ph_{\max}$	Condition	Relay
3.69	1.11e-1	8.07e-2	3.88	3.69	Internal	Trip

Table 3. Results of classifying for inrush current.

Differential Current					Result	
$I_{AZ,\max(post)}^{diff}$	$I_{BZ,\max(post)}^{diff}$	$I_{CZ,\max(post)}^{diff}$	Ph_{sum}	$Ph_{\max}$	Condition	Relay
3.07e+4	1.64e+2	3.08e+4	6.16e+4	3.08e+4	Inrush	Un-Trip

After the decision algorithm process, the algorithm was employed in order to classify fault condition. Case studies were varied so that the algorithm capability can be verified. The considered system is shown in Figure 2. The total number of the case studies was 520. The results obtained from the algorithm proposed in this paper are shown in Table 4.

Table 4. Summary of results from all simulations.

Fault types	Internal faults		External faults		Inrush current	Normal condition
	HV winding	LV winding	HV side	LV side		
Number of cases studies	162	162	90	90	36	6
Detection accuracy	98.15%	100%	87.77%	98.76%	83.33%	100%

5. Conclusion

This paper proposed a technique for detecting and discriminating among inrush current, external fault and internal fault. The simulations, analysis and diagnosis were performed using ATP/EMTP and MATLAB/Simulink. The current waveforms obtained from ATP/EMTP were extracted to several scales with the DWT, and the coefficients of the first scale from the DWT were investigated.

The division between the maximum coefficient of different current phase (A, B, C) and zero sequence different current, was performed as comparison indicators in order to discriminate among inrush current, external fault and internal fault. The results obtained from the algorithm proposed in this paper can detect and indicate the fault condition with the accuracy higher than 96% as presented in Table 4.

Acknowledgments

The authors wish to gratefully acknowledge financial support for this research sponsored by King Mongkut's Institute of Technology Ladkrabang (KMITL), Thailand.

References

1. Z. Bo, G.Weller, T. Lomas, "A New Technique for Transformer Protection Based on Transient Detection," *IEEE Trans. Power Delivery,* 15, 870–875 (2000).
2. T. S. Sidhu, H. S. Gill, M. S. Sachdev, "A Numerical Technique Based on Symmetrical Components for Protecting Three-Winding Transformers," *Electric Power Systems Research,* 54, 19–28 (2000).
3. A. L. Orille-Fernandez, N. K. I. Ghonaim, J. A. Valencia, "A FIRANN as a Differential Relay for Three-Phase Power Transformer Protection," *IEEE Trans. on Power Delivery,* 16, 215–218 (2001).
4. A. Guzman, S. Zocholl, G. Benmouyal, H. J. Altuve, "A Current-Based Solution for Transformer Differential Protection II. Relay Description and Evaluation," IEEE Trans. on Power Delivery, 17, 886–893 (2002).
5. O. A. S. Youssef, "Discrimination Between Faults and Magnetizing Inrush Currents in Transformers Based on Wavelet Transforms," *Electric Power Systems Research,* 63, 87–94 (2002).
6. H. Zhang, P. Liu, O. P. Malik, "A New Scheme for Inrush Identification in Transformer Protection," *Electric Power Systems Research,* 63, 81–86 (2002).
7. Myong-Chul Shin; Chul-Won Park; Jong-Hyung Kim, "Fuzzy Logic-Based Relaying for Large Power Transformer Protection," *IEEE Trans. on Power Delivery,* 18, 718–724 (2003).
8. M. E. Hamedani Golshan, M. Saghaian-nejad, A. Saha, H. Samet, "A New Method for Recognizing Internal Faults from Inrush Current Conditions in Digital Differential Protection of Power Transformers," *Electric Power Systems Research,* 71, 61–71 (2004).
9. G. Baoming, A.T. De'Almeida, Q. Zheng, X. Wang, "An Equivalent Instantaneous Inductance-based Technique for Discrimination Between

Inrush Current and Internal Faults in Power Transformers," *IEEE Trans. on Power Delivery,* 20, 2473–2482 (2005).

10. A. Guzman, H. Altuve, D. Tziouvaras, "Power Transformer Protection Improvements with Numerical Relays," *CIGRE Study Committee B5—Protection and Automation,* 2005.

11. A. Ngaopitakkul, A.Kunakorn and I.Ngamroo, "Discrimination between External Short Circuits and Internal Faults in Transformer Windings using Discrete Wavelet Transforms," *IEEE Industries Application Society Annual Conference 40th,* 448-452 (2005).

12. P. Bastard, P. Bertrand and M. Meunier, "A transformer model for winding fault studies," *IEEE Trans. on Power Delivery,* 9, 690-699 (1994).

13. IEEE working group 15.08.09, *Modeling and analysis of system transients using digital programs,* (IEEE PES special publication)

14. ABB Thailand, Test report no. 56039.

15. *Switching and Transmission Line Diagram* Electricity Generation Authorisation Thailand, 2002.

16. C. Apisit and A. Ngaopitakkul, "Identification of Fault Types for Underground Cable using Discrete Wavelet transform, *Lecture Notes in Engineering and Computer Science: Proceedings of The International MultiConference of Engineers and Computer Scientists 2010, IMECS 2010,* 2, 1262-1266 (2010).

17. S. Kaitwanidvilai, C. Pothisarn, C. Jettanasen, P. Chiradeja and A. Ngaopitakkul, Discrete Wavelet Transform and Back-propagation Neural Networks Algorithm for Fault Classification in Underground Cable, *Lecture Notes in Engineering and Computer Science: Proceedings of The International MultiConference of Engineers and Computer Scientists 2011, IMECS 2011,* 2, 996-1000 (2011).

18. C. Pothisarn, C. Jettanasen, J. Klomjit and A. Ngaopitakkul, Coefficient Comparison Technique of Discrete Wavelet Transform for Discriminating between External Short Circuit and Internal Winding Fault in Power Transformer, *Lecture Notes in Engineering and Computer Science: Proceedings of The International MultiConference of Engineers and Computer Scientists 2012, IMECS 2012,* 2, 1129-1134 (2012).

CLASSIFICATION OF TEMPORAL CHARACTERISTICS OF EPILEPTIC EEG SUBBANDS BASED ON THE LOCAL MAXIMA

S. JANJARASJITT

Department of Electrical and Electronic Engineering,
Ubon Ratchathani University,
Warin Chamrap, Ubon Ratchathani 34190, Thailand
E-mail: ensupajt@ubu.ac.th

Temporal characteristics of EEG signal provide insight into the state of the brain. In this paper, temporal characteristics of epileptic EEG data associated with different pathological states of the brain, i.e., during an epileptic seizure activity and during a non-seizure period, are examined using two features quantified from *local maxima*, referred to as the number of local maxima and the variance of local maxima intervals. The epileptic EEG data are divided into five spectral subbands including δ, θ, α, β and γ subbands. The computational results show that there are substantial differences between the temporal characteristics of epileptic EEG data during an epileptic seizure activity and during a non-seizure period in any subbands. The temporal characteristics of lower frequency subbands of epileptic EEG data during either an epileptic seizure activity or a non-seizure period are also different as compared to their temporal characteristics in higher frequency subbands. From the linear discriminant analysis, the epileptic EEG data druing an epileptic seizure activity can be considerably classified with classification accuracy of 98.50%, sensitivity of 97.00%, and specificity of 100.00%.

Keywords: Electroencephalogram; Epilepsy; Seizure; Local maxima; Subbands.

1. Introduction

Epilepsy is a common brain disorder in which clusters of neurons signal abnormally.[1] More than 50 million individuals worldwide, about 1% of the world's population are affected by epilepsy.[2] In epilepsy, the normal pattern of neuronal activity is disturbed, causing strange sensations, emotions, and behavior, that sometimes include convulsions, muscle spasms, and loss of consciousness.[1] There are many possible causes for seizures ranging from illness to brain damage to abnormal brain development,[1] and epileptic seizures are manifestations of epilepsy.[3] The electroencephalogram (EEG) is a signal that quantifies the electrical activity of the brain, usually from scalp recordings and is commonly used to assess and detect brain abnormalities, and is crucial for the diagnosis of epilepsy.[1]

Temporal patterns of EEG signals can provide important information and such features can be obtained by visual inspection/analysis and using computational tools. Concepts and computational tools derived from the study of complex systems including nonlinear dynamics have gained increasing interest for applications in biology and medicine.[4] The correlation integral and dimension are common nonlinear dynamical analysis techniques that have been applied to EEG signal analysis[5] to study various aspects. Epilepsy is an important application for nonlinear EEG analysis.[6,7]

Local minima and maxima are ones of the most distinguishing temporal characteristics of signals. There are a variety of different approaches to determining local minima and maxima and typically, local min-max detection algorithms rely on thresholding the magnitudes in a specified time window. In Refs. 8–10, the temporal characteristics of epileptic EEG data were examined using features based on the *local minima* and the *local maxima* where the local minima and maxima are defined as points whose amplitudes are less and greater than their neighbors, respectively. Furthermore, in Ref. 11, the temporal characteristics of five spectral subbands of epileptic EEG data were examined using the same features based on the *local minima* and the *local maxima* instead of the full spectral band. The computational results showed that the temporal characteristics of epileptic EEG data during an epileptic seizure activity are substantially different from that during a non-seizure period.[11]

In this study, the temporal characteristics of same five spectral subbands of epileptic EEG data are further examined using the features based on *local maxima*. This reduces the computational time and memory compared to that using the features based on both *local minima* and *local maxima* without changing any temporal characteristic measures. Five spectral subbands of epileptic EEG data examined in this study consist of δ (1–4 Hz), θ (4–8 Hz), α (8–13 Hz), β (14–30 Hz), and γ (31–80 Hz). Two temporal features of the local maxima referred to as the number local maxima N_λ and the variance of local maxima intervals V_λ are examined.

From the computational results, it is shown that the number of local maxima N_λ of epileptic EEG data during an epileptic seizure activity is significantly different from that during a non-seizure period in δ, α, β and γ subbands while the variance of local maxima V_λ of epileptic EEG data during an epileptic seizure activity is sigficantly different from that during a non-seizure period in all five subbands. Also, the epileptic EEG data

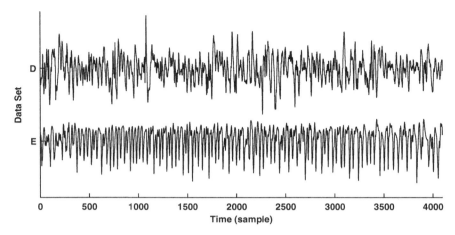

Fig. 1. Examples of intracranial EEG signals of the data sets D (non-seizure) and E (seizure).

during an epileptic seizure activity and during a non-seizure period exhibit different temporal characteristics, i.e., the number of local maxima and the variance of local maxima intervals, at the lower frequency subbands as compared to the higher frequency subbands. Furthermore, the epileptic EEG data during an epileptic seizure activity are considerably classified from the epileptic EEG data during a non-seizure period using the number of local maxima N_λ of the α and γ subbands of epileptic EEG data as the features. The classification accuracy, the sensitivity, and the specificity are 98.50%, 97.00%, and 100.00%, respectively.

2. Methods

2.1. *Data and Subjects*

The intracranial EEG data of epilepsy patients examined in this study were obtained from the Department of Epileptology, University of Bonn (available online at `http://epileptologie-bonn.de/cms/front_content.php?idcat=193&lang=3&changelang=3`) and originated from the study presented in.[12] Two sets of EEG data, referred to as sets D and E, were recorded using intracranial electrodes from five epilepsy patients. Both sets of EEG data were recorded from within the epileptogenic zone. Further, the EEG data in the set D corresponds to non-seizure periods while the EEG data in the set E were recorded during seizure activity.

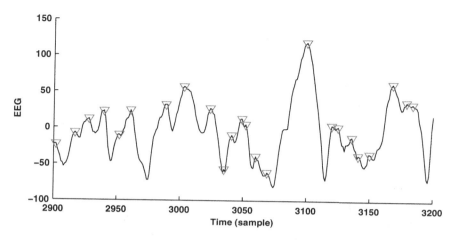

Fig. 2.   Local maxima λ_{max} of an intracranial EEG signal plotted in '$\triangledown$'.

Each epileptic EEG data set contains 100 epochs of a single-channel that were selected to be free of artifacts such as muscle activity and eye movements. The length of each epoch is 23.6 seconds (4097 samples). In addition, the epochs of the EEG signal satisfied the weak stationarity criterion given in.[12] The sampling rate of the data is 173.61 Hz and a bandpass filter with passband between 0.50 Hz and 85 Hz was used during signal acquisition. Examples of the epileptic EEG signals for each data set are depicted in Fig. 1.

2.2. *The Local Maxima and the Temporal Characteristics*

Let the sequence $\{x[n]\}$ be samples from a signal for $n = 0, 1, \ldots, N - 1$, where N is the length of signal. The local maxima λ_{max} are defined as points whose amplitude is greater than that of its consecutive preceding and succeeding points. Mathematically, the local maxima λ_{max} are therefore given by

$$\lambda_{\mathrm{max}} = \left\{ n = \left\lceil \frac{s+t}{2} \right\rceil \, \middle| \, x[s-1] < x[n] \text{ and } x[t+1] < x[n] \right\} \qquad (1)$$

where $x[s] = x[s+1] = \ldots = x[n] = x[n+1] = \ldots = x[t]$. The local maxima of an EEG signal are depicted in Fig. 2.

Temporal characteristics of the spectral subbands of EEG signals are examined based on two features quantified from the local maxima λ_{max} defined as follows:

- *Number of local maxima:*
  The number of local maxima N_λ is the total number of local maxima λ_{max} of the EEG signal.
- *Variance of local maxima intervals:*
  The variance of local maxima intervals V_λ is the variance of the sequence of distances between two consecutive local maxima.

2.3. *Analytical Framework*

In this study, intracranial EEG signals of data sets D and E are divided into five spectral subbands: δ (1–4 Hz), θ (4–8 Hz), α (8–13 Hz), β (14–30 Hz), and γ (31–80 Hz). The temporal characteristics, i.e., the number of local maxima N_λ and the variance of local maxima intervals V_λ, of all five spectral subbands including the δ, θ, α, β and γ subbands are examined. The two-sample t–test is performed to determine whether both local maxima characteristics of each spectral subbands of EEG data set D significantly differs from that of the data set E.

Furthermore, linear discriminant analysis is used to investigate differences between the temporal characteristics of epileptic EEG data during an epileptic seizure activity and that during a non-seizure period. The performance of classification between two sets of epileptic EEG data, i.e., ones during an epileptic seizure activity (set E) and ones during a non-seizure period (set D), is determined from the classification accuracy R, sensitivity S_{sens} and specificity S_{spec}:

$$R = \frac{\#\mathrm{TP} + \#\mathrm{TN}}{\#\mathrm{TP} + \#\mathrm{TN} + \#\mathrm{FP} + \#\mathrm{FN}} \tag{2}$$

$$S_{\mathrm{sens}} = \frac{\#\mathrm{TP}}{\#\mathrm{TP} + \#\mathrm{FN}} \tag{3}$$

$$S_{\mathrm{spec}} = \frac{\#\mathrm{TN}}{\#\mathrm{TN} + \#\mathrm{FP}} \tag{4}$$

where $\#\mathrm{TP}$, $\#\mathrm{TN}$, $\#\mathrm{FP}$, and $\#\mathrm{FN}$ denote the number of epileptic EEG of data set E classified as data set E, the number of epileptic EEG of data set D classified as data set D, the number of epileptic EEG of data set D classified as data set E, and the number of epileptic EEG of data set E classified as data set D, respectively.

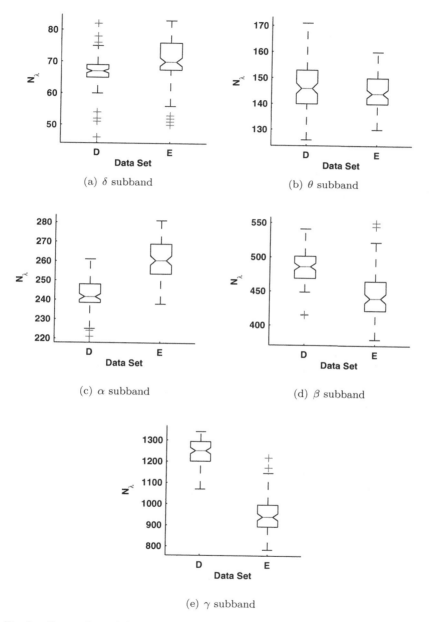

Fig. 3. Comparison of the number of local maxima N_λ of the spectral subbands of the EEG data of sets D and E.

50

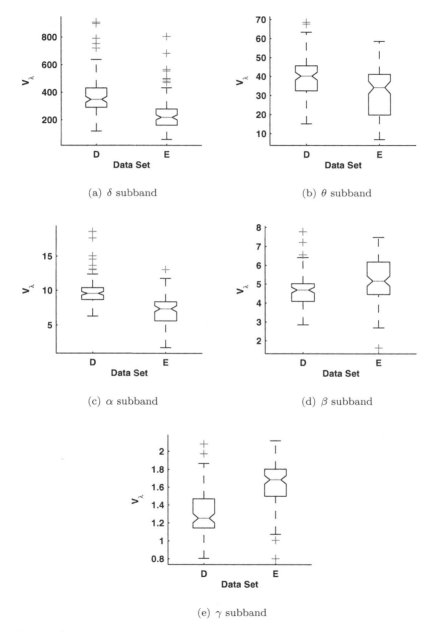

(a) δ subband

(b) θ subband

(c) α subband

(d) β subband

(e) γ subband

Fig. 4.   Comparison of the variance of local maxima intervals V_λ of the spectral subbands of the EEG data of sets D and E.

Table 1. Statistical values of the local maxima characteristics of the spectral subbands of the EEG data of sets D and E.

Feature	Data Set	Subband	Mean	Median	S. D.
N_λ	D	δ	66.6000	67.0000	4.9298
N_λ	D	θ	146.9000	146.0000	10.2725
N_λ	D	α	241.9700	241.5000	7.7283
N_λ	D	β	485.3100	486.0000	23.5907
N_λ	D	γ	1238.9600	1251.0000	65.5126
N_λ	E	δ	70.1600	70.0000	7.3109
N_λ	E	θ	145.0200	144.0000	6.4573
N_λ	E	α	260.3500	260.5000	9.3251
N_λ	E	β	444.7000	440.0000	35.4659
N_λ	E	γ	951.7600	941.5000	80.4724
V_λ	D	δ	368.4740	346.6896	144.8281
V_λ	D	θ	39.8915	40.1767	11.1968
V_λ	D	α	9.7732	9.5612	1.9207
V_λ	D	β	4.6854	4.6762	0.8564
V_λ	D	γ	1.3191	1.2522	0.2554
V_λ	E	δ	243.8120	217.3027	130.7759
V_λ	E	θ	31.5279	34.1755	13.9692
V_λ	E	α	7.0478	7.3763	2.1122
V_λ	E	β	5.2881	5.1693	1.2032
V_λ	E	γ	1.6346	1.6845	0.2641

3. Results

The number of local maxima N_λ of the δ, θ, α, β and γ subbands of epileptic EEG data of sets D and E are, respectively, compared in Fig. 3(a), Fig. 3(b), Fig. 3(c), Fig. 3(d) and Fig. 3(e) while the box plots shown in Fig. 4(a), Fig. 4(b), Fig. 4(c), Fig. 4(d) and Fig. 4(e) compare the variance of local maxima intervals V_λ of the δ, θ, α, β and γ subbands of epileptic EEG data of sets D and E, respectively. Remark that the central mark of each box is the median mark while the edges of the box are the 25th and 75th percentiles. In addition, Table 1 summarizes the mean, median and standard deviation of both the number of local maxima N_λ and the variance of local maxima intervals V_λ of the epileptic EEG subbands.

Evidently, the epileptic EEG data of sets D and E associate with different temporal characteristics based on the local maxima features of various spectral subbands including δ, θ, α, β and γ are substantially different. The number of local maxima N_λ of the δ and α subbands of epileptic EEG data of set E tends to be higher than that of set D. On the contrary, the number of local maxima N_λ of the β and γ subbands of epileptic EEG data of set E tends to be less than that of set D. Also, the variance of local maxima

Table 2.   Results of t-test of the local maxima characteristics of the spectral subbands of the EEG data of sets D and E.

Feature	Subband	Hypothesis	p-value
N_λ	δ	H_0 can be rejected	$p \ll 0.0001$
N_λ	θ	H_0 cannot be rejected	$p = 0.1229$
N_λ	α	H_0 can be rejected	$p \ll 0.0001$
N_λ	β	H_0 can be rejected	$p \ll 0.0001$
N_λ	γ	H_0 can be rejected	$p \ll 0.0001$
V_λ	δ	H_0 can be rejected	$p \ll 0.0001$
V_λ	θ	H_0 can be rejected	$p \ll 0.0001$
V_λ	α	H_0 can be rejected	$p \ll 0.0001$
V_λ	β	H_0 can be rejected	$p \ll 0.0001$
V_λ	γ	H_0 can be rejected	$p \ll 0.0001$

intervals V_λ of the δ, θ and α subbands of epileptic EEG data of set E tends to be lower than that of set D but the variance of local maxima intervals V_λ of the β and γ subbands of epileptic EEG data of set E tends to be higher than that of set D.

Furthermore, the results of the two-sample t-test comparing between the local maxima characteristics, i.e., N_λ and V_λ of the δ, θ, α, β and γ subbands of epileptic EEG data of sets D and E are summarized in Table 2. From the two-sample t-test for the number of local maxima N_λ, the results suggest that there are statistically significant differences between the number of local maxima N_λ of the δ, α, β and γ subbands of epileptic EEG data of sets D and E with a p-value of 0.0001. Similarly, the results of two-sample t-test for the variance of local maxima intervals V_λ suggest that there are statistically significant differences between the variance of local maxima intervals V_λ of epileptic EEG data of sets D and E with a p-value of 0.0001 in all five spectral subbands, i.e., δ, θ, α, β and γ.

Because the number of local maxima N_λ in the α and γ subbands of epileptic EEG data are two temporal characteristics based on the local maxima that have the farthest distance between the epileptic EEG data sets D and E, both are chosen as features used for the classification. From the linear discriminant analysis, the epileptic EEG data of sets D and E can be classified as shown in Fig. 5. The number of epileptic EEG of data set E classified as data set E (#TP), the number of epileptic EEG of data set D classified as data set D (#TN), the number of epileptic EEG of data set D classified as data set E (#FP), and the number of epileptic EEG of data set E classified as data set D (#FN) are 97, 100, 0, and 3, respectively. As a result, the classification accuracy (R), the sensitivity (S_{sens}), and the

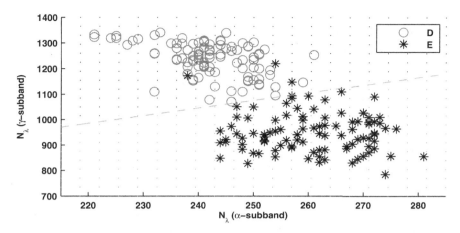

Fig. 5. Classification of the epileptic EEG of data sets D and E using the number of local maxima N_λ in the α and γ subbands.

specificity (S_{spec}) are, respectively, 98.50%, 97.00%, and 100.00%.

4. Conclusions and Discussion

In this study, the temporal characteristics of the spectral subbands of epileptic EEG data as quantified from the local maxima are examined. This is an extended study from Ref. 11 where the temporal characteristics of the spectral subbands of epileptic EEG data quantified from both of the local minima and maxima. This reduces the computational time and memory. There are five spectral subbands examined in this study include δ, θ, α, β, and γ subbands. Also, two temporal characteristics of the spectral subbands of the epileptic EEG data examined are referred to as the number of local maxima N_λ and the variance of local maxima intervals V_λ.

It was observed that the number of local min-max of the full-spectrum EEG signal during an epileptic seizure activity tends to be lower than that during a non-seizure period while the variance of local min-max intervals of the full-spectrum EEG signal during an epileptic seizure activity tends to be higher than that during a non-seizure period.[9,10] Moreover, a decrease of the number of local min-max N_λ and an increase of the variance of local min-max intervals V_λ of the full-spectrum long-term ECoG signal during a seizure onset were observed.[8] Similar to the temporal characteristics of the spectral subbands of the epileptic EEG data during an epileptic seizure activity and during a non-seizure period based on the local minima and maxima, the temporal characteristics of the epileptic EEG data during an epileptic seizure activity and during a non-seizure period based on the local maxima in each subband are distinguishing.

At the lower frequency subbands (i.e., δ and α) the number of local maxima N_λ of epileptic EEG data of set D tends to be higher than that of set E but the number of local maxima N_λ of epileptic EEG data of set D at the higher frequency subbands (i.e., β and γ) tends to be lower than that of set E. This indicates that there is a higher rate of amplitude variation of the low frequency contents of EEG signal during an epileptic seizure activity than that during a non-seizure period. On the other hand, the rate of amplitude variation of the high frequency contents of EEG signal during a non-seizure period tends to be higher than that during an epileptic seizure activity.

In addition, the variance of local maxima intervals V_λ of epileptic EEG data of set D at the lower frequency subbands (i.e., δ, θ and α) tends to be lower than that of set E but at the higher frequency subbands (i.e., β and γ) the variance of local maxima intervals V_λ of epileptic EEG data of set D tends to be higher than that of set E. This indicates that the regularity of amplitude variation of the low frequency contents of the epileptic EEG signal during an epileptic seizure activity is lower than that during a non-seizure period. The regularity of amplitude variation of the high frequency contents of the epileptic EEG signal during an epileptic seizure activity is higher than that during a non-seizure period.

Therefore, as quantified using the local minima and maxima, at the low frequency subbands the rate of amplitude variation of epileptic EEG signal during an epileptic seizure activity is higher but its temporal variability is less as compared to that during a non-seizure period. This is however opposite at the high frequency subbands. The rate of amplitude variation of epileptic EEG signal during an epileptic seizure activity is lower while the regularity of amplitude variation is less as compared to that during a non-seizure period.

Also, the temporal characteristics of epileptic EEG data, i.e., the number of local maxima N_λ and the variance of local maxima intervals V_λ, can be used for classification of epileptic EEG data during an epileptic seizure activity. The best classification can be obtained using the number of local maxima N_λ of the α and γ subbands of epileptic EEG data with classification accuracy of 98.50%, sensitivity of 97.00%, and specificity of 100.00% for this specific sets of epileptic EEG data. This thus suggests that simple computational features such as the number of local maxima N_λ can be used for epileptic seizure detection and classification.

Acknowledgments

This research is supported by Ubon Ratchathani University.

References

1. National Institute of Neurological Disorders and Stroke (NINDS), Seizure and epilepsy: Hope through research (2004), `http://www.ninds.nih.gov/disorders/epilepsy/detail_epilepsy.htm`.
2. B. Litt and J. Echauz, *Lancet Neurology* **1**, 22 (2002).
3. A. Subasi, *Expert Systems with Applications* **29**, 343 (2005).
4. A. L. Goldberger, *Proc. Am. Thorac. Soc.* **3**, 467 (2006).
5. W. S. Pritchard and D. W. Duke, *Brain Cogn.* **27**, 353 (1995).
6. C. E. Elger, G. Widman, R. Andrzejak, J. Arnhold, P. David and K. Lehnertz, *Epilepsia* **41(Suppl.)**, S34 (2000).
7. C. E. Elger, G. Widman, R. Andrzejak, M. Dumpelman, J. Arnhold, P. Grassberger and K. Lehnertz, *Value of nonlinear time series analysis of the EEG in neocortical epilepsies*, in *Advance in Neurology*, eds. P. D. Williamson, A. M. Siegel, D. W. Roberts, V. M. Thadani and M. S. Gazzaniga (Lippincott Williams & Wilkins, 2000).
8. S. Janjarasjitt and K. A. Loparo, Temporal variability of the ECoG signal during epileptic seizures, in *The 2010 ECTI International Conference on Electrical Engineering/Electronics, Computer, Telecommunications and Information Technology*, (Chiang Mai, Thailand, 2010).
9. S. Janjarasjitt and K. A. Loparo, Comparison of temporal variability of epileptic ECoG signals, in *Proceedings of 2010 International Conference on Electronics and Information Engineering*, (Kyoto, Japan, 2010).
10. S. Janjarasjitt and K. A. Loparo, Investigation of temporal variability of epileptic eeg signals, in *Proceedings of IEEE Region 10 Annual International Conference*, (Fukuoka, Japan, 2010).
11. S. Janjarasjitt and K. A. Loparo, Examination of temporal characteristics of epileptic EEG subbands based on the local min-max, *Lecture Notes in Engineering and Computer Science: Proceedings of The International Multi-Conference of Engineers and Computer Scientists 2012*, IMECS 2012, 14–16 March 2012, Hong Kong, pp. 1095–1099.
12. R. G. Andrzejak, K. Lehnertz, F. Mormann, C. Rieke, P. David and C. E. Elger, *Phys. Rev. E* **64**, 061907 (2001).

A CONCURRENT ERROR DETECTION AND CORRECTION BASED FAULT-TOLERANT XOR-XNOR CIRCUIT FOR HIGHLY RELIABLE APPLICATIONS[*]

MOUNA KARMANI

*Electronics & Microelectronics Laboratory, Monastir, Tunis University, Tunisia.
Email: mouna.karmani@yahoo.fr*

CHIRAZ KHEDHIRI

*Electronics & Microelectronics Laboratory, Monastir, Tunis University, Tunisia.
Email: chirazkhedhiri@yahoo.fr*

BELGACEM HAMDI

*Electronics & Microelectronics Laboratory, Monastir, Tunis University and ISSAT,
Sousse, Tunisia. Email: Belgacem.Hamdi@issatgb.rnu.tn*

KA LOK MAN

*Xi'an Jiaotong-Liverpool University, China, Myongji University, South Korea and Baltic
Institute of Advanced Technology, Lithuania. Email: ka.man@xjtlu.edu.cn*

ENG GEE LIM

Xi'an Jiaotong-Liverpool University, China. Email: enggee.lim@xjtlu.edu.cn

CHI-UN LEI

University of Hong Kong, Hong Kong. Email: culei@eee.hku.hk

The continuous shrinking of MOS devices dimensions affects the circuit performance and reliability by introducing transient and permanent faults that can cause critical failures which make Fault-tolerant VLSI integrated circuits a typical requirement especially in safety-critical applications. Thus, in this work, we address the issue of fault-tolerant chip based on using Concurrent Error Detection and Correction architectures. Since XOR-XNOR circuits are basic building blocks in various digital and mixed systems, especially in arithmetic circuits, these gates should be designed such that they indicate and correct any malfunction during normal operation. In fact, the property of verifying the results

---

[*]This research work is partially sponsored by SOLARI (HK) CO (www.solari-hk.com) and KATRI
(www.katri.co.jp and www.katri.com.hk).

delivered by a circuit during its normal operation is called Concurrent Error Detection (CED) while the property of correcting a considered fault during normal operation is called Concurrent Error Correction (CEC). In this work, we propose a concurrent error detection and correction based fault-tolerant XOR-XNOR circuit implementation. The proposed design is implemented using the 32 nm process technology.

1. Introduction

MOS transistor scaling has been the key of the rapid advances of integrated circuits (ICs) performance and density [1]. As technology advances to deep sub-micron levels and below, VLSI circuits increase in complexity and become more susceptible to process variations [2]. The primary effect of process variations is on transistor parameters. Thus, parameter variations in key device parameters such as channel length, threshold voltage and oxide thickness, are increasing at an alarming rate [3]. Due to these parameter variations in VLSI circuits, transient and permanent faults arise; and they can corrupt the circuit operation. Thus, fault-tolerant designs are required to ensure safe operation of digital systems performing safety-critical functions in safety-critical devices [4]. Actually, the introduction of fault tolerance gives the possibility to accept circuits containing some failures and thus achieve better manufacture yield. The fault tolerance property gives also solution to correct the failures occurring when using the chip which is very interesting because it increases the availability of the considered device [5].

To achieve the fault-tolerance property, it is important to increase the level of error detection. Thus, Concurrent Error Detection (CED) is important in highly dependable computing systems, because CED techniques can be used to detect permanent and transient faults in these circuits during normal operation [6]. Thereby, in applications where dependability is important, CED circuitry must be used for assuring early detection of errors preserving the state of the system and preventing data corruption [7]. The concurrent error correction (CEC) is also another important property in high reliable systems in which it is not feasible to repair (as in computers on board satellites) or in which the computer is serving a critical function and cannot be lost even for the duration of a replacement (as insight control computers on an aircraft) or in which the repair is prohibitively expensive [8]. In fact, by using the CEC property, since a permanent or transient fault is detected it will be immediately corrected in real-time during the normal operation of the considered system. The Triple Modular Redundancy (TMR) method is a typical error correction hardware scheme. However, its disadvantage is its high hardware redundancy (it needs at least 200% of hardware redundancy) [9]. The CEC based fault-tolerant methods used in older structures have mainly been designed for coping with situations of single faults. However, multiple faults are expected to occur in circuits because of the increasing failure rate and therefore the fault tolerance methods capable of handling multiple faults are also strongly needed [5].

The exclusive-OR (XOR) and exclusive-NOR (XNOR) are fundamental components in full adders, and in larger circuits such as parity checkers. Thus, the performance of these logic circuits is affected by the individual performance of each XOR-XNOR included in them [10-11]. In this work, we propose a fault-tolerant XOR-XNOR design with concurrent error detection and correction capabilities. The circuit implementation is achieved using the 32 nm process technology.

This XOR-XNOR circuit implementation is proposed to achieve the required level of reliability and robustness for schemes using the dual duplication code like adders, ALUs, multipliers and dividers. Simulation results of the implemented chip are presented and show that the technique is effective and can be easily implemented in the System-on- Chip (SoC) environment. We first present our proposed circuit topology and its simulation results (Section 2). Then we verify the proposed circuit using three typical fault models (Section 3). Finally the proposed Concurrent error correction architecture is illustrated in Section 4.

2. Concurrent Error Detection and Correction Based Fault-Tolerant Systems

As a result of advances in technology shrinking device dimensions, complementary metal oxide semiconductor (CMOS) chips manufacturing process are moved to the nano regime. Therefore there are new problems that appear and at the same time a list of old problems are getting more severe [5]. In fact, process variation is caused by the inability to precisely control the fabrication process at small-feature technologies [12]. Therefore, deep-submicron technologies with lower voltage level systems are more susceptible to permanent and transient faults. Consequently, fault tolerance must be used to tolerate design faults in safety-critical systems. Thus, building fault-tolerant systems is so important for safety-critical applications (such as transport and medical applications) to ensure the correctness of the results computed in the presence of permanent and transient faults [4-7].

For safety-critical applications, the correspondent safety level requires the detection and correction of any single fault that occurs during normal operation. In order to ensure this on-line fault detection property, we can employ CED techniques. The most basic method of performing CED is hardware redundancy, i.e., two copies of the hardware are used concurrently to perform the same computation on the same data. At the end of each computation, the results are compared and any discrepancy is reported as an error [13]. In fact, the CED technique presented in this work is achieved by means of output duplication technique. The output of a circuit has a certain property that can be monitored by a checker. If an error causes a violation of the property, the checker gives an error indication signal [6]. Actually, in safety-critical systems and especially when it is not possible to achieve manual repairs in real time the CEC property is more than indispensable to achieve the nodded level of reliability and performance. Thus, in order to ensure the concurrent error correction property in our XOR-XNOR circuit implementation we will add adopted hardware architecture based on using a very simple network of two inputs CMOS multiplexer circuits.

Exclusive-OR (XOR) and exclusive-NOR (XNOR) circuits are basic building blocks in various digital systems, especially in arithmetic circuits. Also, the performance of these logic circuits is affected by the individual performance of each XOR-XNOR circuit included in them. Thus, each XOR-XNOR gate included in these circuits must be fault-tolerant to be able to continue operating even with failures in their hardware [14]. XOR and XNOR circuits implement functions that are complementary. XOR and XNOR circuits are binary operations that perform the following Boolean functions [14]:

$$\begin{cases} A \text{ XOR } B = A \oplus B = A{\sim}B + AB{\sim} \\ A \text{ XNOR } B = A \ominus B = AB + A{\sim}B{\sim} \end{cases}$$

Where (A, A~) and (B, B~) are complementary pairs of data.

The XOR and XNOR circuits can be implemented in different architectures by using different circuit designs. Examples of design techniques for XOR-XNOR circuits are static CMOS logic, pass transistor logic, CMOS pass transistor logic, double pass transistor logic and transmission gate [14-15].

Pass transistor logic uses fewer transistors to implement important logic functions. Also, smaller transistors and smaller capacitances are required, and it is faster than conventional CMOS. However, the pass transistor gates generate degraded signals, which slow down signal propagation [16].

2.1. *The proposed XOR-XNOR circuit implementation*

In this work, a novel XOR-XNOR circuit designed in modified pass transistor logic is presented in Figure 1. The current implementation does not generate

60

degraded signals. This gate has dual inputs (A, A~, B and B~) and generates duplicated dual outputs (XOR1, XNOR1) and (XOR2, XNOR2). The circuit implementation is performed with eight MOS transistors [17].

In the current XOR-XNOR circuit, the fault-tolerance property is ensured by using a duplicated output computation based concurrent error detection method. In fact, this CED method is based on generating duplicated dual outputs. The first path gives the first outputs (XOR1 and XNOR1); and the second path gives the second outputs (XOR2 and XNOR2). Errors caused by faults will affect only one of the two paths and may be detected just by checking the complementarity principle between each (XOR, XNOR) function. The proposed XOR-XNOR circuit and the correspondent layout are respectively given by Figure 1 and Figure 2.

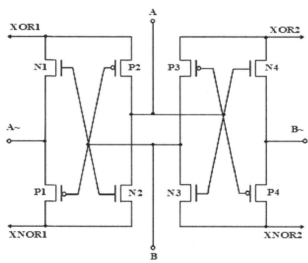

Figure 1. The proposed XOR-XNOR circuit implementation with duplicated output computation.

The XOR and XNOR functions are both performed using two different paths. From the first path we obtain XOR1 and XNOR1 functions (1). In this path the inputs are A, A~, B and B~, but the input B~ is performed from the input B. Thus, the circuit will be insensitive to any kind of errors affecting the input B~.

$$\begin{cases} \text{XOR1}= \text{A XOR B}=\text{A\~B} + \text{A}(\textbf{B\~}) \\ \text{XNOR1}=\text{A XNOR B}= \text{AB} + \text{A\~}(\textbf{B\~}) \end{cases} \quad (1)$$

From the second path, we obtain XOR2 and XNOR2 functions (2). In this path, the inputs are A, A~, B and B~, but the input A~ is performed from the input A. Thus, the circuit will be insensitive to any kind of errors affecting the input A~.

$$\begin{cases} \text{XOR2= A XOR B=(A~)B + AB~} \\ \\ \text{XNOR2=A XNOR B= AB + (A~)B~} \end{cases} \tag{2}$$

Thus, this XOR-XNOR circuit implementation can increase the fault tolerance property, since the circuit outputs are computed using the output computation of two paths.

2.2. Simulation results

The XOR-XNOR circuit is implemented in full-custom 32 nm technology [14]. SPICE simulations of the circuit extracted from the layout, including parasitic, are used to demonstrate that the circuit has a conformed electrical behaviour.

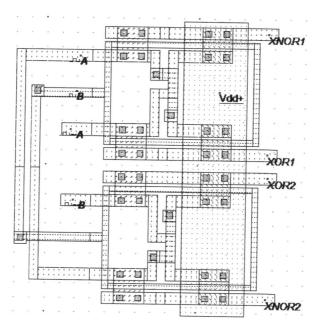

Figure 2. Layout of the XOR-XNOR circuit in full-custom 32 nm process technology.

SPICE simulations of the circuit without any fault are illustrated by Figure 3.

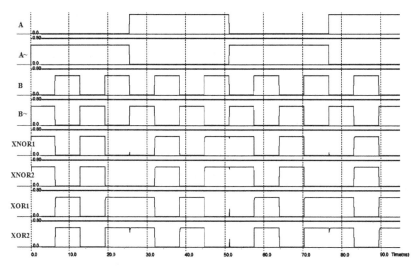

Figure 3. SPICE simulations of the XOR-XNOR circuit in 32nm technology without faults.

From this simulation we can remark that the outputs (XOR1, XNOR1) and (XOR2, XNOR2) obtained by this computation technique are complementary, therefore the circuit is fault-free. Also, as indicated in the previous section, the current implementation does not generate degraded output signals and can produce strong 1's and 0's. This is important, especially with low voltage levels and small noise margins.

3. The XOR-XNOR Circuit Fault Analysis

Due to the diversity of VLSI defects, it is hard to generate complementary tests for real defects. Therefore, fault models are necessary to analyse any VLSI circuit in the presence of faults. In the following sub-sections, we analyse the behaviour of our XOR-XNOR circuit with respect to the set of fault models including logical stuck-at faults, transistor stuck-on and transistor stuck-open faults.

3.1. The stuck-at fault model

The most common model used for logical faults is the single stuck-at fault. It assumes that a fault in a logic gate results in one of its inputs or the output is fixed at either a logic 0 (stuck-at-0) or at logic 1 (stuck-at-1) [15]. So, for inputs, we consider the logical stuck-at fault model. Table 1 gives the response of the gate for all inputs combinations.

Table 1. The Gate Response for all inputs combinations.

A	A~	B	B~	xor1	xnor1	xor2	xnor2	Conclusion
0	0	0	0	0	0	0	0	Multiple fault (detected)
0	0	0	1	0	0	0	1	Single fault (detected & corrected)
0	0	1	0	0	0	1	0	Single fault (detected & corrected)
0	0	1	1	0	0	1	1	Multiple fault (detected)
0	1	0	0	0	1	0	0	Single fault (detected & corrected)
0	1	0	1	0	1	0	1	Valid input
0	1	1	0	1	0	1	0	Valid input
0	1	1	1	1	0	1	1	Single fault (detected & corrected)
1	0	0	0	1	0	0	0	Single fault (detected & corrected)
1	0	0	1	1	0	1	0	Valid input
1	0	1	0	0	1	0	1	Valid input
1	0	1	1	0	1	1	1	Single fault (detected & corrected)
1	1	0	0	1	1	0	0	Multiple fault (detected)
1	1	0	1	1	1	1	0	Single fault (detected&corrected)
1	1	1	0	1	1	0	1	Single fault (detected&corrected)
1	1	1	1	1	1	1	1	Multiple fault (detected)

From the table above, we can conclude that for primary logical stuck-at faults, all single and multiple faults on primary inputs will result in a non-valid code by producing no complementary outputs. In other words, each fault will be detected when there are non complementary (XOR, XNOR) outputs, because normally XOR and XNOR should be complementary data. We should note that the error detection is achieved by using only one of the two paths. Also, the fault-free outputs are available on the second path. (The concurrent error correction is available only for single stuck-at faults).

However, not all defects in VLSI circuits can be represented by the stuck-at fault model. It has been shown that transistor stuck-on and transistor stuck-open are two other types of defects that may remain undetected if testing is performed only based on the stuck-at fault model [16-19]. Next, we consider the stuck-on and stuck-open transistor fault model. We will examine all possible single transistor stuck-on and transistor stuck-open faults within the circuit of Figure 1 in next two sub-sections.

3.2. *The transistor stuck-on fault model*

A stuck-on transistor fault involves the permanent closing of the path between the source and the drain of the transistor (PMOS or NMOS). In other words, a transistor stuck-on fault may be modelled as a bridging fault from the source to the drain of a transistor [19]. In order to analyse the circuit behaviour in the presence of stuck-on faults with realistic circuit defects, we simulate the considered XOR-XNOR circuit in the presence of faults. Faults are manually injected in the circuit layout of Figure 2. Table 2 states the circuit response for all possible single transistor stuck-on faults.

Table 2. The Gate Response for Transistor Stuck-on faults.

Transistor Stuck-on	Input vector detecting the fault A A~ B B~	XOR1	XNOR1	XOR2	XNOR2	EI1	EI2
N1	1 0 0 1	0	0	1	0	1	0
N2	0 1 0 1	0	0	0	1	1	0
N3	0 1 0 1	0	1	0	0	0	1
N4	0 1 1 0	1	0	0	0	0	1
P1	1 0 1 0	0	0	0	1	1	0
P2	0 1 1 0	0	0	1	0	1	0
P3	1 0 0 1	1	0	0	0	0	1
P4	1 0 1 0	0	1	0	0	0	1

The two signals EI1 (Error indication 1) and EI2 (Error indication 2) are obtained by checking the principle of complementarity respectively between

(XOR1, XNOR1) and (XOR2, XNOR2). EI1 and EI2 are generated using respectively the first path and the second path outputs computation.

If a fault appears, it only affects one of the two paths. Consequently, a fault producing no complementary outputs affects only one of the two error indication signals (EI1 and EI2). Each error detection signal can be generated using a pass transistor XNOR gate.

3.3. *The transistor stuck-on fault model*

A stuck-open transistor involves the permanent opening of the connection between the source and the drain of a transistor [19]. When a transistor is rendered non-conducting by a fault, it is said to be stuck-open. In our fault model, a single physical line in the circuit is broken. In fact, by examining the layout of the circuit given by Figure 1, we can remark that transistors N1, N2, P1 and P2 have the same gate which is connected to the input B. Also, transistors N3, N4, P3 and P4 have the same gate which is connected to the input A. The transistors gates for each output block are connected in such a way that a single break in any transistor gate does not make the transistor stuck-open. Therefore, we need two breaks to make any transistor stuck-open. Thus, this property makes the circuit fault-tolerant for single stuck-open fault model. In this section, we have shown that the scheme of the Figure 1 is fault-tolerant for the logic stuck-at fault model, transistor stuck-on and stuck-open fault model.

4. The Proposed Concurrent Error Correction Architecture

The proposed CEC architecture is illustrated by Figure 4. The considered architecture uses two error indication circuits which are Pass Transistors XNOR gates. In fact, the two circuits EI1 and EI2 are used to indicate any violation of the complementarity principle respectively for (XOR1, XNOR1) and (XOR2, XNOR2). The output of each error indication circuit is the bit address of a 2 inputs (2 to 1) CMOS multiplexer circuit. Thus, the first multiplexer circuit is used to output the fault-free XOR output (XOR1 or XOR2) while the second one is used to output the fault-free XNOR output (XNOR1 or XNOR2).

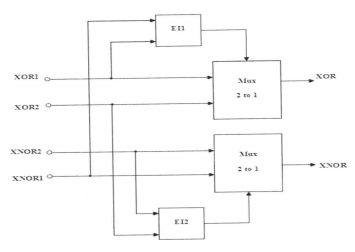

Figure 4. The proposed CEC architecture.

Table 3 illustrates the CEC circuit response for all possible combination of the two error indication circuit outputs.

Table 3. The CEC circuit response for all possible combination of the two error indication circuit outputs.

EI1	EI2	Conclusion	XOR	XNOR
0	0	The two paths are fault-free	XOR1	XNOR2
0	1	The path 1 is fault-free	XOR1	XNOR1
1	0	The path 2 is fault-free	XOR2	XNOR2
1	1	The two paths are faulty	XOR2	XNOR1

From the table above we can conclude that the concurrent error correction architecture illustrated by Figure 4 corrects only single faults. In fact, when multiple faults occur (the last case) the two paths can produce faulty outputs. For safety critical application requiring a high level of reliability all cases must be taken in consideration. Therefore, to obtain a better level of reliability we can further complicate the circuit in order to avoid the faulty outputs obtained when multiple faults occur. The next proposed architecture uses only the logic product EI1*EI2 signal and the two multiplexed outputs (XOR and XNOR). Table 4 presents all cases obtained when multiple faults occur. In fact XOR_{FF} and $XNOR_{FF}$ are the final XOR-XNOR circuit outputs which are supposed to be fault-free even with presence of single or multiple faults. The XOR_{FF} and

XNOR$_{FF}$ values are given by the design engineer who can theatrically predict the needed values to be transmitted when multiple faults occur.

Table 4. The cases obtained when multiple faults occur.

XOR1 XNOR1 XOR2 XNOR2	EI1*EI2	XOR XNOR	XOR$_{FF}$ XNOR$_{FF}$
0000	1	0 0	0 1
0011	1	1 0	1 0
1100	1	0 1	1 0
1111	1	1 1	0 1

From the table above and using the KARGNAUX map we can conclude that when a multiple fault occur (EI1*EI2 is equal to the high level) the fault-free multiplexed XOR and XNOR outputs are respectively given by:

$$\begin{cases} XOR_{FF} = XOR \oplus XNOR \\ XNOR_{FF} = XOR \odot XNOR \end{cases}$$

The additional concurrent error correction architecture used to correct multiple faults is illustrated by Figure 5.

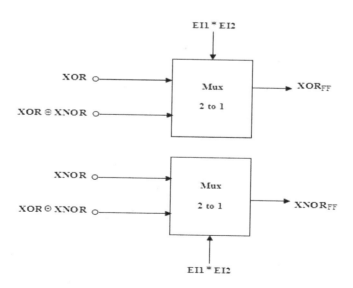

Figure 5. The proposed additional multiple fault CEC architecture.

The logical product EI1*EI2 is the bit address of the two used multiplexers. Thus, when there is no multiple faults (EI1*EI2 is equal to the low level) the multiplexed outputs are XOR and XNOR. Otherwise, when multiple faults occurs (EI1*EI2 is equal to the high level) the multiplexed outputs are XOR⊕XNOR and XOR⊖XNOR.

5. Conclusion

With the continuous scaling of devices and interconnects, the geometrical feature size decreases from submicron to the sub-nano and beyond. Therefore, process variations become relatively more important and VLSI complex integrated circuits are more susceptible to permanent and transient faults. Therefore, designing fault-tolerant systems providing continuous and safe operation in the presence of faults become important, especially in specific applications domains requiring very high levels of reliability. In this work, we have presented a Concurrent Error Detection and Correction based fault-tolerant XOR-XNOR circuit implementation. The proposed circuit can significantly improves the reliability and robustness for schemes using the dual duplication code such as adders, ALUs, multipliers and dividers. The fault tolerance property ensured by using this XOR-XNOR circuit makes the design very efficient and suitable for real-time VLSI applications.

References

1. M. White and Y. Chen, "Scaled CMOS Technology Reliability Users Guide", *NASA Electronic Parts and Packaging (NEPP) Program*, (2008).
2. M. Orshansky, S. R. Nassif and D. Boning, "Design for manufacturability and statistical design: A constructive approach", *US Springer,* pp. 1–8 (2008).
3. R. Garg and S. P. Khatri, "Analysis and Design of Resilient VLSI Circuits: Mitigating Soft Errors and Process Variations", *US Springer,* pp. 1-10 (2010).
4. E. F. Hitt and D. Mulcar, "Fault-Tolerant Avionic", *CRC Press LL* (2001).
5. Tejio Lehtonen, Juha Plosila and Jouni Isoaho, "On fault tolerance methods for networks-on-chip", *TUCS Technical Report*, (2005).
6. C. Zeng and E. J. McCluskey, "Finite State Machine Synthesis with Concurrent Error Detection", *Proc. International Test Conference*, pp. 672-679 (1999).
7. D. Das and N. A. Touba, "Synthesis of Circuits with Low-Cost Concurrent Error Detection Based on Bose-Lin Codes", *Journal of Electronic Testing: Theory and Applications*, Vol. 15, Nos. 1/2, pp. 145-155 (1999).

8. VS Veeravalli, "Fault tolerance for arithmetic and logic unit", *IEEE Southeast con* 09 329–334 (2009).

9. Hussain Al-Asaad and Edward Czeck, "Concurrent error correction in iterative circuits by recomputing with partitioning and voting", *Digest of Papers of the Eleventh Annual 1993 IEEE VLSI Test Symposium,* pp. 174-177 (1993).

10. M. Nicolaidis, R. O. Duarte, S. Manich, and J. Figueras, "Fault-Secure Parity Prediction Arithmetic Operators", *In IEEE Design & Test of computers,* Vol. 14, pp. 60-71 (1997).

11. R. Chowdhury, A. Banerjee, A. Roy and H. Saha, "A High Speed Transistor Full Adder Design using Novel 3 Transistor XOR Gates", *In International Journal of Electronics, Circuits and Systems II,* pp. 217-223 (2008).

12. S. R. Sarangi, B. Greskamp, R. Teodorescu, J. Nakano, A. Tiwari, and J. Torrellas, "VARIUS: A model of process variation and resulting timing errors for microarchitects", *In IEEE Transactions on Semiconductor Manufacturing,* vol. 21, pp. 3-13 (2008).

13. N. Joshi, K. Wu, J. Sundararajan, and R. Karri, "Concurrent Error Detection for Evolutional Functions with applications in Fault Tolerant Cryptographic Hardware Design", *IEEE Transactions on Computer-Aided Design of Integrated Circuits and Systems,* vol. 25, pp. 1163–1169 (2006).

14. H. Mishra, S. Wairya, R. K. Nagaria, and S. Tiwari, "New Design Methodologies for High Speed Low Power XOR-XNOR Circuits", *World Academy of Science, Engineering and Technology,* vol. 55, pp. 200-206 (2009).

15. S. Mishra, A. Kumar and R.K. Nagari, "A comparative performance analysis of various CMOS design techniques for XOR and XNOR circuits", *International Journal on Emerging Technologies,* vol. 1, pp. 1-10 (2010).

16. B. Hamdi, C. Khedhiri, and R. Tourki, "Pass Transistor Based Self-Checking Full Adder". *International Journal of Computer Theory and Engineering,* Vol. 3, No. 5, pp. 608-616 (2011).

17. Mouna Karmani, Chiraz Khedhiri, Belgacem Hamdi, Ka Lok Man, Eng Gee Lim and Chi-Un Lei, "A Concurrent Error Detection Based Fault-Tolerant 32 nm XOR-XNOR Circuit Implementation", *Lecture Notes in Engineering and Computer Science: Proceedings of The International MultiConference of Engineers and Computer Scientists 2012, IMECS 2012, 14-16 March, 2012, Hong Kong.*

18. E. Sicard, "Microwind and Dsch version 3.1", *INSA Toulouse,* ISBN 2-87649-050-1, Dec 2006.

19. P.K. Lala, "An introduction to logic circuit testing", *Morgan & Claypool,* pp. 1–9 (2009).

PROBABILITY DISTRIBUTIONS ON AN AND-OR TREE
UNDER DIRECTIONAL ALGORITHMS

TOSHIO SUZUKI[i]* and RYOTA NAKAMURA[ii]†

Department of Mathematics and Information Sciences,
Tokyo Metropolitan University,
Minami-Ohsawa, Hachioji, Tokyo 192-0397, Japan.
[i] *E-mail: toshio-suzuki@tmu.ac.jp*
[ii] *E-mail: r-nakamura@shimz.co.jp*

Consider a probability distribution d on the truth assignments to a perfect binary AND-OR tree. Liu and Tanaka (2007) extends the work of Saks and Wigderson (1986), and they characterize the eigen-distribution, the distribution achieving the equilibrium, as the uniform distribution on the 1-set (the set of all reluctant assignments for which the root has the value 1). We show that the uniqueness of the eigen-distribution fails provided that we restrict ourselves to directional algorithms. An alpha-beta pruning algorithm is said to be directional (Pearl, 1980) if for some linear ordering of the leaves (Boolean variables) it never selects for examination a leaf situated to the left of a previously examined leaf. We also show that the following weak version of the Liu-Tanaka result holds for the situation where only directional algorithms are considered; a distribution is eigen if and only if it is a distribution on the 1-set such that the cost does not depend on an associated deterministic algorithm.

Keywords: AND-OR tree; directional algorithm; computational complexity; randomized algorithms.

1. Introduction

A Boolean decision tree[1] is a kind of algorithm widely used in diagnosis systems. Given a Boolean function $f(x, y, z)$, an algorithm A is a Boolean decision tree of f if A works in the following manner. At the beginning, the three bits x, y, z of a given input are covered. Each of them has the value 1 or 0. First, A chose one of bits, say y, and uncovers it. Then, A may return the value of $f(x, y, z)$. Or, A choses one of the remaining two bits, say z, and repeats the above procedure.

---

*This work was supported in part by JSPS KAKENHI (C) 22540146 and (B) 23340020.
† Current affiliation is Shimizu Corporation.

Algorithms for an AND-OR tree is a typical example of a decision tree. An AND-OR tree is a tree whose internal nodes are labeled either AND ($\wedge$) or OR ($\vee$). The concept of an AND-OR tree is interesting because of its two aspects, a Boolean function and a game tree. In this paper, we use the terminology in more restricted sense. An *AND-OR tree* (an *OR-AND tree*, respectively) denotes a tree such that its root is an AND-gate (an OR-gate), layers of AND-gates and those of OR-gates alternate and each leaf is assigned Boolean value. 1 denotes true and 0 denotes false. A perfect binary tree of this type with height $2k$ is denoted by T_2^k.

Here, note that we consider two kinds of trees, a tree probed and a tree probes. An AND-OR tree is probed. It is a special case of a game tree where the range of an evaluation function is $\{0, 1\}$. Thus, a MIN-gate in this case is an AND-gate, and a MAX-gate is an OR-gate. On the other hand, a game-tree search algorithm probes an AND-OR tree, where each leaf of an AND-OR tree is considered as a bit of a given input. Thus, an algorithm finding the root value of an AND-OR tree is a special case of a decision tree. The cost of the computation is measured by the number of leaves probed. In an advanced setting, we consider a probability distribution on the truth assignments. Then, the concept of equilibrium is naturally defined.

There are classical results showing that we may restrict ourselves to algorithms of a particular type when we investigate an AND-OR tree. For example, see Refs. 7,12. We consider alpha-beta pruning algorithms only. An alpha-beta pruning algorithm is a special case of a depth-first mini-max algorithm. An alpha-beta pruning algorithm is characterized by the following two properties; whenever the algorithm knows a child node of an AND-gate has the value 0, the algorithm recognizes that the AND-gate has the value 0 without probing the other child node (such a saving of cost is said to be an alpha-cut), and whenever the algorithm knows a child node of an OR-gate has the value 1, the algorithm recognizes that the OR-gate has the value 1 without probing the other child node (a beta-cut). See Ref. 3 for more on alpha-beta pruning algorithms.

Now, we review three important previous results. The first one is Yao's principle. It is a variant of Von-Neumann's Min-Max theorem. In this context, a randomized algorithm denotes a probability distribution on a set of deterministic algorithms. This is a kind of Las-Vegas algorithm. For a randomized algorithm, the cost is defined as to be the expected value of the cost. A probability distribution on the truth assignments is easier to handle than a probability distribution on the algorithms.

Yao's principle[1,14] says that the randomized complexity equals to the distributional complexity.

$$\min_{A_R} \max_{\omega} \mathrm{cost}(A_R, \omega) = \max_{d} \min_{A_D} \mathrm{cost}(A_D, d)$$

The left-hand side is the *randomized complexity*. It is the minimum of maximum of the cost, where the maximum is taken among all truth assignments, and the minimum is taken among all randomized algorithm. The right-hand side is the *distributional complexity*. It is the maximum of minimum of the cost, where the minimum is taken among all deterministic algorithm and the maximum is taken among all probability distribution of truth assignments.

The second important previous result is Saks-Wigderson theorem.[8] Saks and Wigderson establish basic results on the randomized complexity. In particular, they show that the randomized complexity is exponential order of the height of the tree, and they explicitly find the base of exponential. And, they conjecture that a similar estimation holds for any Boolean function.

We ask which distribution achives the equlibrium. The third important previous result is Liu-Tanaka theorem.[4] Liu and Tanaka extend the work of Saks and Wigderson, and characterize a probability distribution achieving the equilibrium of T_2^k. In particular, they show that such a distribution is unique. A distribution d_0 on the truth assignments is the *eigen-distribution* if it achieves the distributional complexity, that is:

$$\min_{A_D} \mathrm{cost}(A_D, d_0) = \max_{d} \min_{A_D} \mathrm{cost}(A_D, d)$$

To be more precise, by extending the concept of a reluctant input in the paper of Saks and Wigderson, Liu and Tanaka define the concept of *i-set* (for $i \in \{0, 1\}$) as the set of all assignments such that the root has the value i and whenever an AND-node has the value 0 (and, whenever an OR node has the value 1), its one child node has the value 1 and the other child node has the value 0. They define an E^i-*distribution* as to be a distribution on the i-set such that all the deterministic algorithms have the same cost. They prove that, for a probability distribution d on the truth assignments to the leaves of T_2^k, the followings (LT1)–(LT3) are equivalent.

(LT1) d is the eigen-distribution;

(LT2) d is an E^1-distribution;

(LT3) d is the uniform distribution on the 1-set.

The current paper is motivated by an example that "contradicts to" the uniqueness of the eigen-distribution. We show that the uniqueness of the eigen-distribution fails when we restrict ourselves to directional algorithms (section 4). In particular, there are uncountably many eigen-distributions in the setting. The goal of the current paper is to extend the Liu-Tanaka theorem to the case where only directional algorithms are considered.

Here, directional algorithm is defined as follows. In general, an algorithm can move in such a way that, if a leaf X is skipped then probe Y before probing Z, otherwise probe X, next probe Z before probing Y. Thus, the priority of Y and Z depends on the history of computation. An algorithm is said to be *directional* (Pearl[6]) if the priority does not depend on the history. We also show that the following weak version of the result of Liu and Tanaka holds in the above setting; a distribution is eigen if and only if E^1 (section 3).

Our main method for showing this equivalence is the no-free-lunch theorem (NFLT, for short). Wolpert and MacReady[2,13] show that, under certain assumptions, averaged over all cost functions, all search algorithms give the same performance. Intuitively, if an algorithm performs well for a certain cost function, then the algorithm performs badly for a certain other cost function. This result is known as NFLT. For a family of algorithms closed under transposition of sub-trees, a variant of NFLT holds, and NFLT implies the equivalence.

The current paper is a revised version of our conference paper.[10] The journal paper[11] is an extended version of Ref. 10.

2. Notation

We denote the empty string by λ. By $\{0,1\}^n$, we denote the set of all strings of length n. The cardinality of a set X is denoted by $|X|$. For sets X and Y, $X - Y$ denotes $\{x \in X : x \notin Y\}$. We let prob$[E]$ denote the probability of an event E. A *k-round AND-OR tree* denotes a perfect binary AND-OR tree of height $2k$ ($k \geq 1$), and is denoted by T_2^k. The subscript 2 denotes that it is a binary tree. For example, T_2^1 is as in Fig. 1. For a positive integer k, replace each leaf of T_2^k by T_2^1; Then, the resulting tree is T_2^{k+1}.

Convention Throughout the paper, unless specified, h denotes a positive integer and T denotes a perfect binary tree of height h such that T is either an AND-OR tree or an OR-AND tree. $\mathcal{A}_D$ denotes the family of all deterministic alpha-beta pruning algorithms calculating the root-value of T. For the definition of *an alpha-beta pruning algorithm*, see Introduction. $\mathcal{A}$ denotes a non-empty subset of $\mathcal{A}_D$. The height h is fixed in the definition of $\mathcal{A}_D$. Thus, we should write, for example, $\mathcal{A}_D(h)$ in the precise manner, but we omit h. The same remark will apply to $\mathcal{A}_{\text{dir}}$ in Definition 2.2. Ω is a non-empty family of assignment-codes, where we define assignment-codes in the following. We label each node of T by a string as follows.

Definition 2.1. A *node-code* is a binary string of length at most h. Each node of T is assigned a node-code in such a way that the code of the root is the empty string, and nodes with codes of the form $u0, u1$ are child nodes of the node with code u. See Fig. 2. A node-code is a *leaf-code* if its length is h. Otherwise, it is an *internal node-code*. An *assignment-code* is a function of the leaf-codes to $\{0, 1\}$.

Fig. 1. T_2^1

Fig. 2. Node-codes

By $\mathcal{A}_{\mathrm{dir}}$, we denote the family of all directional algorithms in $\mathcal{A}_D$. Formal definitions are as follows.

Definition 2.2. Suppose that A_D is a member of the $\mathcal{A}_D$. Let $\ell = 2^h$.

(1) Suppose that $\langle u^{(1)}, u^{(2)}, \cdots, u^{(m)} \rangle$ is a sequence of strings, ω is an assignment-code and that the followings hold.

    (a) $m(\leq \ell)$ is the total number of leaves queried during the computation of A_D under the assignment ω.

    (b) For each $j(1 \leq j \leq m)$, the j-th query in the computation of A_D under the assignment ω is the leaf of code $u^{(j)}$.

    By "*query-history of* $\langle A_D, \omega \rangle$" and "*answer-history of* $\langle A_D, \omega \rangle$", we denote $\langle u^{(1)}, u^{(2)}, \cdots, u^{(m)} \rangle$ and $\langle \omega(u^{(1)}), \cdots, \omega(u^{(m)}) \rangle$, respectively.

(2) $\mathcal{A}_{\mathrm{dir}}$ is a family of algorithms defined as follows. A deterministic algorithm $A_D \in \mathcal{A}_D$ belongs to $\mathcal{A}_{\mathrm{dir}}$ if there exists a permutation $\langle v^{(1)}, v^{(2)}, \cdots, v^{(\ell)} \rangle$ of the leaves such that for every assignment code ω, the query-history of $\langle A_D, \omega \rangle$ is consistent with the permutation. More precisely, the query history is either equal to the permutation, or a sequence (say, $\langle v^{(1)}, v^{(3)}, v^{(4)} \rangle$) given by omitting some leaf-codes from the permutation.

For $A_D \in \mathcal{A}_D$ and an assignment-code ω, we let $C(A_D, \omega)$ denote the number of leaves scanned in the computation of A_D under ω. By the phrase "the cost of A_D with respect to ω", we denote (not time-complexity but) $C(A_D, \omega)$. If d is a probability distribution on the assignments then $C(A_D, d)$ denotes the expected value of the cost with respect to d. For the definitions of the 0-*set* and the 1-*set*, see Introduction.

Definition 2.3. Suppose that $\mathcal{A}$ is a non-empty subset of $\mathcal{A}_D$, and that Ω is a non-empty set of assignment-codes.

(1) A distribution d on Ω is called an *eigen-distribution with respect to* $\langle \mathcal{A}, \Omega \rangle$ if the following holds.

$$\min_{A_D \in \mathcal{A}} C(A_D, d) = \max_{d'} \min_{A_D \in \mathcal{A}} C(A_D, d'),$$

where d' runs over all probability distributions on Ω. In the case where Ω is the set of all assignment-codes, we say "d is an eigen-distribution with respect to $\mathcal{A}$".

(2) Let $i \in \{0, 1\}$. A distribution d on the i-set is called an E^i-*distribution with respect to* $\mathcal{A}$ if there exists a real number c such that for every $A_D \in \mathcal{A}$, it holds that $C(A_D, d) = c$.

(3) (Liu and Tanaka[4]) A distribution d on the truth assignments is eigen (respectively, E^0, E^1) if it is so with respect to $\mathcal{A}_D$.

3. The equivalence of eigen and E^1

We show that the equivalence of eigen and E^1 holds even when we restrict ourselves to directional algorithms. We develop a framework including both the directional case and the usual case.

Our main method for showing this equivalence is the no-free-lunch theorem (NFLT, for short). See Introduction for the informal explanation on NFLT. We apply NFLT to a class closed under transposition. Before stating formal definitions, we give an informal explanation.

Suppose that u is an internal node. Just below the node u, there are two sub-trees. Suppose that we transpose the role of the two. Then, by using a natural isomorphism between the two sub-trees, we can naturally define the concept of u-transposition of a leaf. A node in the left sub-tree corresponds to another node in the right sub-tree in such a way that the leftmost leaf in the left sub-tree corresponds to the leftmost leaf in the right sub-tree. If a leaf y is not a descendant of u, we define its u-transposition as to be y itself. In the same way, we can define the u-transposition of a truth assignment, and the u-transposition of an algorithm. Formal definitions are as follows.

Definition 3.1. Suppose that u is an internal node-code.

(1) Suppose v and v' are node-codes of the same length. We say "v' *is the* u-*transposition of* v" (in symbol, $v' = \mathrm{tp}_u(v)$) if one of the below holds.

   (a) There exist $i \in \{0, 1\}$ and a string w such that $v = uiw$ (concatenation) and $v' = u(1 - i)w$.

   (b) u is not a prefix of v and it holds that $v = v'$.

(2) Suppose that ω, ω' are assignment-codes. We say "ω' *is the* u-*transposition of* ω" (in symbol, $\omega' = \mathrm{tp}_u(\omega)$) if $\omega'(v) = \omega(\mathrm{tp}_u(v))$ holds for each leaf-code v.

(3) Suppose that A_D and A'_D are deterministic algorithms. We say "A'_D *is the* u-*transposition of* A_D" (in symbol, $A'_D = \mathrm{tp}_u(A_D)$) if the following holds: "For each assignment-code ω, the query-history of $\langle A'_D, \omega \rangle$ is given by applying component-wise tp_u operation to the query-history of $\langle A_D, \mathrm{tp}_u(\omega) \rangle$." To be more precise, denote the query-history of $\langle A_D, \mathrm{tp}_u(\omega) \rangle$ and that of $\langle A'_D, \omega \rangle$ by $\langle x^{(1)}, \cdots, x^{(m)} \rangle$ and $\langle y^{(1)}, \cdots, y^{(m)} \rangle$, respectively. Then, the following holds.

$$\forall j \leq m \ \ y^{(j)} = \mathrm{tp}_u(x^{(j)})$$

And, the answer history of $\langle A_D, \mathrm{tp}_u(\omega) \rangle$ is the same as that of $\langle A'_D, \omega \rangle$.

Example 3.1. We consider the case where $h = 2$. Then the followings hold. $\mathrm{tp}_\lambda(abcd) = cdab$, $\mathrm{tp}_0(abcd) = bacd$ and $\mathrm{tp}_1(abcd) = abdc$, where we denote a truth assignment ω by a string $\omega(00)\omega(01)\omega(10)\omega(11)$.

Definition 3.2.

(1) $\mathcal{A}$ is *closed (under transposition)* if for each $A_D \in \mathcal{A}$ and for each internal node-code u, we have $\mathrm{tp}_u(A_D) \in \mathcal{A}$.

(2) Ω is *closed (under transposition)* if for each $\omega \in \Omega$ and for each internal node-code u, we have $\mathrm{tp}_u(\omega) \in \Omega$.

(3) Ω is *connected (with respect to transposition)* if for every distinct members $\omega, \omega' \in \Omega$, there exists a finite sequence $\langle \omega_i \rangle_{i=1,\cdots,N}$ in Ω and a finite sequence $\langle u^{(i)} \rangle_{i=1,\cdots,N-1}$ of strings such that $\omega_1 = \omega, \omega_N = \omega'$ and for each $i < N$, ω_{i+1} is the $u^{(i)}$-transposition of ω_i.

Convention Throughout the rest of the section, $\mathcal{A}$ denotes a non-empty closed subset of $\mathcal{A}_D$.

Definition 3.3. Suppose that $p_1, \cdots, p_n$ are non-negative real numbers such that their sum makes 1. And, suppose that $\Omega_1, \cdots, \Omega_n$ are mutually disjoint non-empty families of assignment-codes. In addition, suppose that $d_1, \cdots, d_n$ are distributions such that each d_j is a distribution on Ω_j.

(1) $p_1 d_1 + \cdots + p_n d_n$ denotes the distribution d on $\Omega_1 \cup \cdots \cup \Omega_n$ defined as follows. For each j ($1 \leq j \leq n$) and each truth assignment $\omega \in \Omega_j$, we have prob[d is ω] $= p_j \times$ prob[d_j is ω].

(2) Given a distribution d, we say "d is a distribution on $p_1 \Omega_1 + \cdots + p_n \Omega_n$" if there exist distributions d_j' on Ω_j ($1 \leq j \leq n$) such that $d = p_1 d_1' + \cdots + p_n d_n'$.

The following is a variant of the no-free-lunch theorem.

Lemma 3.1. *Suppose $p_1, \cdots, p_n$ and $\Omega_1, \cdots, \Omega_n$ satisfy the requirements in Definition 3.3, and each Ω_j is connected. Then, there exits a real number c such that for every distribution d on $p_1 \Omega_1 + \cdots + p_n \Omega_n$, the following holds.*

$$\sum_{A_D \in \mathcal{A}} C(A_D, d) = c \tag{1}$$

Proof. We investigate the case of $n = 1$. The general case is immediately shown by the case of $n = 1$. For every assignment-code ω and for every internal node-code u, the mapping of $A_D \in \mathcal{A}$ to $\mathrm{tp}_u(A_D)$ is a permutation of $\mathcal{A}$. And, we have $C(\mathrm{tp}_u(A_D), \omega) = C(A_D, \mathrm{tp}_u(\omega))$. Hence, the sum of $C(A_D, \omega)$ over all $A_D \in \mathcal{A}$ is the following.

$$\sum_{A_D \in \mathcal{A}} C(\mathrm{tp}_u(A_D), \omega) = \sum_{A_D \in \mathcal{A}} C(A_D, \mathrm{tp}_u(\omega)) \tag{2}$$

Therefore, there exists a real number c such that for every $\omega \in \Omega$, $\sum_{A_D \in \mathcal{A}} C(A_D, \omega) = c$. Hence, the left-hand side of (1) is equal to the following.

$$\sum_{A_D \in \mathcal{A}} \sum_{\omega \in \Omega} \mathrm{prob}[d = \omega] C(A_D, \omega) \sum_{\omega \in \Omega} \left(\mathrm{prob}[d = \omega] \sum_{A_D \in \mathcal{A}} C(A_D, \omega) \right) = c \tag{3}$$

$\square$

Lemma 3.2. *Suppose that $p_1, \cdots, p_n$ and $\Omega_1, \cdots, \Omega_n$ satisfy the requirements in Definition 3.3. And, suppose that each Ω_j is closed.*

(1) Let $d_{\mathrm{unif.}}(p_1 \Omega_1 + \cdots + p_n \Omega_n)$ denote the distribution $p_1 d_1 + \cdots + p_n d_n$, where each d_j is the uniform distribution on Ω_j. Then, there exits a real number c such that for every deterministic algorithm $A_D \in \mathcal{A}_D$, it holds that $C(A_D, d_{\mathrm{unif.}}(p_1 \Omega_1 + \cdots + p_n \Omega_n)) = c$.

(2) Suppose that each Ω_j is not only closed but also connected and that d is a distribution on $p_1 \Omega_1 + \cdots + p_n \Omega_n$. Then, the following (a), (b) and (c) are equivalent, where B_j are any deterministic algorithms (not necessarily in $\mathcal{A}$), and $d_{\mathrm{unif.}}(\Omega_j)$ is the uniform distribution on Ω_j.

(a) *The following holds, where d' runs over distributions on $p_1\Omega_1 + \cdots + p_n\Omega_n$.*

$$\min_{A_D \in \mathcal{A}} C(A_D, d) = \max_{d'} \min_{A_D \in \mathcal{A}} C(A_D, d') \qquad (4)$$

(b) *There exits a real number c such that for every $A_D \in \mathcal{A}$, it holds that $C(A_D, d) = c$.*

(c)

$$\min_{A_D \in \mathcal{A}} C(A_D, d) = \sum_{j=1}^{n} p_j C(B_j, d_{\mathrm{unif.}}(\Omega_j)) \qquad (5)$$

Proof. 1. In the same way as the proof of Lemma 3.1, the case of $n \geq 2$ is reduced to the case of $n = 1$. Thus, in the following, we prove the case of $n = 1$ by induction on h. The case of $h = 1$ is immediate. At the the induction step, let T_0 (T_1, respectively) be the left (right) sub-tree just under the root. Since Ω_1 is closed, the assertion is equivalent to its weaker form: "The costs are the same for all algorithms which probe T_0 before T_1." We call such algorithms *"left-first algorithms"* in this proof.

Now, Ω_1 is partitioned into sets such that each component Ω' is of the following form. There exist strings α_0 and α_1 (depending on Ω') such that α_i is an assignment-code for T_i for each i, and the component Ω' is the direct product of the closure of α_0 and that of α_1. Here, the closure of α_i denotes the following set.

$$\{\mathrm{tp}_u(\alpha_i) : \ u \text{ is an internal node-code}\} \qquad (6)$$

By the induction hypothesis, the cost of a left-first algorithm depends only on Ω', and does not depend on an algorithm. Hence, the same holds with respect to Ω_1.

2. By the assertion 1 of the current lemma and Lemma 3.1, each of the assertions (a) and (b) is equivalent to (7).

$$\min_{A_D \in \mathcal{A}} C(A_D, d) = \frac{1}{|\mathcal{A}|} \sum_{A_D \in \mathcal{A}} C(A_D, d) \qquad (7)$$

Therefore, by the assertion 1, (a) is equivalent to the following.

$$\min_{A_D \in \mathcal{A}} C(A_D, d) = \min_{A_D \in \mathcal{A}} C(A_D, d_{\mathrm{unif.}}(p_1\Omega_1 + \cdots + p_n\Omega_n)) \qquad (8)$$

And, the right-hand side of (8) equals to the right-hand side of (5). Hence, (a) is equivalent to (c). □

Lemma 3.3. *Assume that T is an AND-OR tree. Suppose that d is an eigen-distribution with respect to $\mathcal{A}$ (see Definition 2.3). Then d is a distribution on the 1-set.*

Proof. For each positive integer h and each $i \in \{0,1\}$, we let $c_i^{\wedge,h}$ ($c_i^{\vee,h}$, respectively) denote $C(A_D, d_{\mathrm{unif.}}(i\text{-set}))$ for the perfect binary AND-OR tree (OR-AND tree, respectively) of height h. Here, A_D is an element of $\mathcal{A}$. By Lemma 3.2, $c_i^{\wedge,h}$ ($c_i^{\vee,h}$) is well-defined regardless of the choice of A_D.

Claim 1. Suppose that Ω is closed. If a given tree is an AND-OR tree and Ω is not the 1-set (a given tree is an OR-AND tree and Ω is not the 0-set, respectively), then for any deterministic algorithm A_D, $C(A_D, d_{\mathrm{unif.}}(\Omega))$ is less than $c_1^{\wedge,h}$ ($c_0^{\vee,h}$, respectively).

Proof of Claim 1 (sketch): By induction on h, the followings are shown.

$$c_0^{\wedge,h} = c_1^{\vee,h} < c_1^{\wedge,h} = c_0^{\vee,h} \leq \frac{4}{3} c_0^{\wedge,h} \tag{9}$$

By means of these inequality, the claim is shown by induction. Q.E.D.(Claim 1)

Now, suppose that T is an AND-OR tree. Suppose that d is an eigen-distribution with respect to $\mathcal{A}$. Let $\langle \Omega_j : j = 1, \cdots, n \rangle$ be a partition of the set of all truth assignments to connected closed sets. Without loss of generality, Ω_1 is the 1-set. For each j, let p_j be the probability of d being a member of Ω_j. By Lemma 3.2, (5) holds. Hence, by Claim 1, there are positive real numbers $c_2, \cdots, c_n$ such that the following holds.

$$\forall j \geq 2 \quad c_j < c_1^{\wedge,h} \tag{10}$$

$$\min_{A_D \in \mathcal{A}} C(A_D, d) = p_1 c_1^{\wedge,h} + \sum_{j=2}^{n} p_j c_j \tag{11}$$

Since d is eigen with respect to $\mathcal{A}$, d achieves the maximum value of (11). Hence, it holds that $p_1 = 1$ and $p_j = 0$ for all $j \geq 2$. Thus, d is a distribution on the 1-set. □

Theorem 3.1. *Assume that a given tree T is T_2^k for some positive integer k. Suppose that a family $\mathcal{A}$ of algorithms is closed under transposition and that d is a probability distribution on the assignment-codes. Then, the followings are equivalent (see Definition 2.3).*

(LT1$^{\mathcal{A}}$) d is an eigen-distribution with respect to $\mathcal{A}$.

(LT2$^{\mathcal{A}}$) d is an E^1-distribution with respect to $\mathcal{A}$.

Proof. By Lemma 3.3, (LT1$^{\mathcal{A}}$) is equivalent to "d is an eigen-distribution with respect to $\langle \mathcal{A}, (1\text{-set})\rangle$". By Lemma 3.2, this is equivalent to (LT2$^{\mathcal{A}}$).

□

4. A case where the uniqueness fails

A direct corollary to Lemma 3.2 is the following.

Corollary 4.1. *Assume that a given tree T is T_2^k for some positive integer k. Then, (LT3) implies (LT2$^{\mathcal{A}}$):*
 (LT3) d is the uniform distribution on the 1-set.
 (LT2$^{\mathcal{A}}$) d is an E^1-distribution with respect to $\mathcal{A}$.

We show that the uniqueness of the eigen-distribution fails in the directional case. This is shown by proving that (LT2$^{\mathcal{A}}$) does not imply (LT3) with respect to $\mathcal{A} = \mathcal{A}_{\text{dir}}^k$ (see below).
    Convention $\mathcal{A}_{\text{dir}}^k$ denotes $\mathcal{A}_{\text{dir}}$ (see Definition 2.2) in the case where $h = 2k$ and $T = T_2^k$.

		A_1	A_2	A_3	A_4	A_5	A_6	A_7	A_8
		1234	4312	3421	2143	3412	1243	2134	4321
ω_1	1010	2	3	3	4	2	3	3	4
ω_2	1001	3	2	4	3	3	2	4	3
ω_3	0110	3	4	2	3	3	4	2	3
ω_4	0101	4	3	3	2	4	3	3	2

Now, we investigate the case of $k = 1$. Table 1 shows the values of $C(A_D, \omega_i)$ for each ω_i in the 1-set. We denote an assignment-code ω by a string $\omega(00)\omega(01)\omega(10)\omega(11)$. And, each A_j is the name of an element of $\mathcal{A}_{\text{dir}}^1$. Recall Definition 2.2. Each A_j is determined by a permutation $xyzw$ of $\{0,1\}^2$ that shows priority of scanning leaves. A string such as 1234 denotes a permutation of the above property, where we denote leaf-codes 00, 01, 10 and 11 by numerals 1, 2, 3 and 4, respectively.

Theorem 4.1.[9]

(1) There are uncountably many E^1-distributions with respect to $\mathcal{A}_{\text{dir}}^1$. Hence, (LT2$^{\mathcal{A}}$) does not imply (LT3) with respect to $\mathcal{A} = \mathcal{A}_{\text{dir}}^1$.
(2) There are uncountably many eigen-distributions with respect to $\mathcal{A}_{\text{dir}}^1$.

Proof. Suppose $0 \leq \varepsilon \leq 1/2$. Let d_ε denote the distribution d on the 1-set such that the probabilities of d being $\omega_1, \omega_2, \omega_3$ and ω_4 are $\varepsilon, 1/2 - \varepsilon, 1/2 - \varepsilon$ and ε, respectively. By Table 1, the value $C(A_j, d_\varepsilon)$ does not depend on j. And, d_ε is not the uniform distribution on the 1-set unless $\varepsilon = 1/4$. Thus, the assertion 1 holds. Hence, by Theorem 3.1, the assertion 2 holds. □

On the other hand, it is easy to see that E^0-distribution with respect to $\mathcal{A}^1_{\text{dir}}$ is unique.

By means of Theorem 4.1 and induction on k, we can show the following.

Theorem 4.2.[9] *For each positive integer k, the statements of Theorem 4.1 hold for $\mathcal{A}^k_{\text{dir}}$ in place of $\mathcal{A}^1_{\text{dir}}$.*

5. A case where the uniqueness holds

In this section, we give an alternative proof for the characterization of the eigen-distribution (in the usual case) as the uniform distribution on the 1-set.[4] To be more precise, we show that (LT2$^\mathcal{A}$) implies (LT3) with respect to the un-directional algorithms (see Corollary 4.1).

An example of an element of $\mathcal{A}^1_D - \mathcal{A}^1_{\text{dir}}$ is as follows. It begins with scanning the leaf of code 00. If a beta-cut does not happen there, the query-history is $\langle 00, 01, 10, 11 \rangle$. Otherwise, the query-history is $\langle 00, 11, 10 \rangle$, where the leaf-code 01 is skipped due to the beta-cut. By taking transpositions of this algorithm, we know that $\mathcal{A}^1_D - \mathcal{A}^1_{\text{dir}}$ consists of 8 algorithms.

Theorem 5.1.[5] *Suppose that $h \geq 2$, where h is the height of T. Then, (LT2$^\mathcal{A}$) implies (LT3) with respect to $\mathcal{A} = \mathcal{A}_D - \mathcal{A}_{\text{dir}}$.*

Proof. For each $i \in \{0, 1\}$ and a positive integer g, let $(i$-set$)^g$ denote the i-set in the case of $h = g$. To be more precise, the i-set for an AND-OR tree and that for an OR-AND tree are different. However, it is easy to see which one is considered, according to the context. By induction on $h \geq 2$, we shall show the following requirement R_h.

R_h: "Suppose that $i \in \{0, 1\}$ and that a distribution d on $(i$-set$)^h$ is an E^i-distribution with respect to $\mathcal{A} = \mathcal{A}_D - \mathcal{A}_{\text{dir}}$. Then, d is the uniform distribution on $(i$-set$)^h$."

The base case R_2 is shown by solving equations; it is in the same way as our proof of the uniqueness of E^0-distribution with respect to $\mathcal{A}^1_{\text{dir}}$.

Suppose that R_n holds. In the rest of the proof, let $h = n + 1$, $i \in \{0, 1\}$ and assume that d is a distribution on $(i$-set$)^{n+1}$ and that d is an E^i-

distribution with respect to $\mathcal{A}_D - \mathcal{A}_{\text{dir}}$. We investigate the case where the root is an AND-gate and $i = 1$. The other cases are shown in the same way.

Let T_0 (T_1, respectively) be the left (right) sub-tree just under the root. For each assignment α on T_0 (such that $\alpha \in (1\text{-set})^n$ and the denominator of (12) is positive), consider the distribution d_α on T_1 as follows. For each assignment β on T_1, we let prob$[\ d_\alpha$ is $\beta\ ]$ as to be the following conditional probability, where $\alpha\beta$ denotes the concatenation of α and β.

$$\text{prob}[\ d \text{ is } \alpha\beta\ |\ \exists x\ d \text{ is } \alpha x\ ] \tag{12}$$

By the induction hypothesis R_n, for all α such that $\alpha \in (1\text{-set})^n$ and the denominator of (12) is positive, d_α is the uniform distribution on $(1\text{-set})^n$. The same holds for the case where the roles of T_0 and T_1 are exchanged.

Now, by the induction hypothesis R_n, it is not hard to see that the requirement R_{n+1} is satisfied. $\qquad\square$

6. Conclusive remarks

By extending the work of Tarsi,[12] it is shown by Saks and Wigderson that the randomized complexity of an AND-OR tree is the same as that for directional algorithms; for more pricise, see Theorem 5.2 of Ref. 8. In the case of T_2^k, by using our results in § 3, the above result is extended as follows.

Proposition 6.1. *Suppose that k is a positive integer and T is T_2^k. And, suppose that $\mathcal{A}$ is a non-empty subset of $\mathcal{A}_D$ and $\mathcal{A}$ is closed under transposition. Then, the following holds, where d runs over all distributions.*

$$\max_d \min_{A_D \in \mathcal{A}_D^k} C(A_D, d) = \max_d \min_{A \in \mathcal{A}} C(A_D, d) \tag{13}$$

Proof. Let $d_{\text{unif.}}$ be the uniform distribution on the 1-set. By Lemma 3.2 and Lemma 3.3, the both sides of (13) are equal to the following.

$$\min_{A_D \in \mathcal{A}_D^k} C(A_D, d_{\text{unif.}}) = \min_{A_D \in \mathcal{A}} C(A_D, d_{\text{unif.}}) \qquad\square$$

Hence, the $\mathcal{A}_D$ (the class of all deterministic algorithms) and $\mathcal{A}_{\text{dir}}$ (that of all directional algorithms) have the same distributional complexity.

In contrast, they do not agree on the question of "Which distribution achieves the equilibrium?" A variant of the no-free-lunch theorem implies the equivalence of "eigen" and "E^1", but it does not imply the uniqueness of the eigen-distribution. The set of all *un*-directional algorithms plays an important role to show the uniqueness.

Acknowledgment

The authors would like to thank M. Kumabe, C.-G. Liu, Y. Niida, K. Ogawa, K. Tanaka and T. Yamazaki for helpful discussions.

References

1. Arora, S. and Barak, B.: *Computational Complexity.* (Cambridge university press, New York, 2009).
2. Ho, Y.C. and Pepyne, D.L.: Simple explanation of the no-free-lunch theorem and its implications. *J. Optimiz. Theory App.*, **115** pp.549–570 (2002).
3. Knuth, D.E. and Moore, R.W.: An analysis of alpha-beta pruning. *Artif. Intell.*, **6** pp. 293–326 (1975).
4. Liu, C.-G. and Tanaka, K.: Eigen-distribution on random assignments for game trees. *Inform. Process. Lett.*, **104** pp.73–77 (2007).
5. Nakamura, R.: Randomized complexity of AND-OR game trees. Master thesis, Department of Mathematics and Information Sciences, Tokyo Metropolitan University, Tokyo (2011).
6. Pearl, J.: Asymptotic properties of minimax trees and game-searching procedures. *Artif. Intell.*, **14** pp.113–138 (1980).
7. Pearl, J.: The solution for the branching factor of the alpha-beta pruning algorithm and its optimality. *Commun. ACM*, **25** pp.559–564 (1982).
8. Saks, M. and Wigderson, A.: Probabilistic Boolean decision threes and the complexity of evaluating game trees. in: *Proc. 27th Annual IEEE Symposium on Foundations of Computer Science (FOCS)*, pp.29–38 (1986).
9. Suzuki, T.: Failure of the uniqueness of eigen-distribution on random assignments for game trees. *Sūrikaisekikenkyūsho-Kokyuroku*, **1729** pp.111–116, Research Institute for Mathematical Sciences, Kyoto University (2011).
10. Suzuki, T. and Nakamura, R.: Probability distributions achieving the equilibrium of an AND-OR tree under directional algorithms. *Lecture Notes in Engineering and Computer Science: Proceedings of The International Multi-Conference of Engineers and Computer Scientists 2012, IMECS 2012*, (14-16 March, 2012, Hong Kong), pp.194-199. http://www.iaeng.org/publication/IMECS2012/
11. Suzuki, T. and Nakamura, R.: The eigen distribution of an AND-OR tree under directional algorithms. *IAENG International Journal of Applied Mathematics*, **42:2**, pp.122-128. http://www.iaeng.org/IJAM/issues_v42/issue_2/index.html
12. Tarsi, M.: Optimal search on some game trees. *J. ACM*, **30** pp. 389–396 (1983).
13. Wolpert, D.H. and MacReady, W.G.: No-free-lunch theorems for search. *Technical report SFI-TR-95-02-010*, (Santa Fe Institute, New Mexico, 1995).
14. Yao, A.C.-C.: Probabilistic computations: towards a unified measure of complexity, in: *Proc. 18th annual IEEE symposium on foundations of computer science (FOCS)*, pp.222–227 (1977).

AN EFFICIENT DIFFERENTIAL FULL ADDER

CHIRAZ KHEDHIRI

Electronic & Microelectronics Laboratory
Monastir, Tunisia, chirazkhedhiri@yahoo.fr

MOUNA KARMANI

Electronic & Microelectronics Laboratory
Monastir, Tunisia, mouna.karmani@yahoo.fr

BELGACEM HAMDI

Electronic & Microelectronics Laboratory
Monastir, Tunisia.
ISSAT, Sousse, belgacem.hamdi@gmail.com

KA LOK MAN

Xi'an Jiaotong-Liverpool University, China and Baltic Institute of
Advanced Technologies, Lithuania, ka.man@xjtlu.edu.cn

In this chapter, an efficient differential full adder is presented. The circuit is simulated in double pass transistor CMOS at 32nm technology. This fully differential adder contains only 20 transistors which mean that we save 66.66% of the transistors number overhead if we compare the proposed design to the duplication based adder.

1. Introduction

Addition is a very basic operation in arithmetic. Subtraction, multiplication, division and address calculation are some of the well-known operations based on addition. In most of these systems the adder is part of the critical path that determines the overall performance of the system. That is why enhancing the performance of the 1-bit full-adder cell is a significant goal [1].

A variety of full-adders using different logic styles and technologies have been reported in literature [2-3-4-5-6-7]. Although all of them perform a similar function, but the method of producing the intermediate nodes and the outputs, the loads on them and the transistor count are varied. Different logic styles tend to favor one performance aspect at the expense of the other.

Some of them use one logic style for the whole full adder and the others use more than one logic style for their implementation [8].

1.1. The CMOS full adder

A complementary static CMOS circuit consists of an NMOS pull-down network connecting the ground to the output and a dual PMOS pull-up network connecting the power to the output [9] as it is shown in Fig.1.

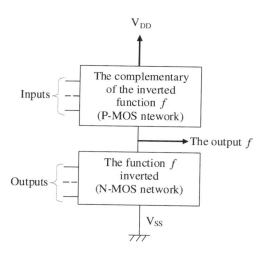

Figure 1. General structure of the static CMOS logic circuit.

The advantage of complementary CMOS style is its robustness against voltage scaling and transistor sizing, which are essential to provide reliable operation at low voltage and arbitrary transistor sizes [10].

A complementary static CMOS adder using 28 transistors is shown in Fig. 2. This adder implements the following Boolean functions:

$$Sum = [\overline{(\overline{a + b} + \overline{c}_{in})c_{out} + \overline{a\,b\,c_{in}}}] \qquad (1)$$

$$Carry = [\overline{\overline{ab} + \overline{(a + b)}c_{in}}] \qquad (2)$$

86

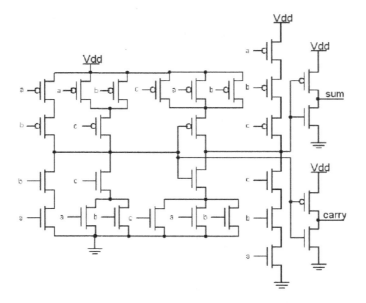

Figure 2. 28T static complementary CMOS full adder [11].

1.2. *The TGA full adder*

A full adder using transmission gates TGA is presented in Figure 3. It contains 20 transistors.

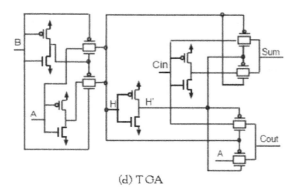

(d) TGA

Figure 3. 20T TGA full adder [12].

1.3. *The Complementary Pass Transistor Logic CPL) full adder*

In 1989, researches from Hitachi Central Research Laboratories in Japan published the structure known as Complementary Pass Transistor Logic (CPL) [13].

The main concept behind CPL is the use of an nMOS pass transistor network for logic organisation, and elimination of the pMOS latch. CPL consists of complementary inputs/outputs, an nMOS pass transistor logic network, and CMOS output inverters. The pass transistors function as pull-down and pull-up devices [14].

The CPL was significant in the fact that it was based on the use of the pass transistor network. The logic function, which is built from the pass transistors, not only efficiently utilises the silicon, but results in a very fast logic which is also characterised by low power consumption [15].

A differential CPL adder using 28 transistors is shown in Fig. 4.

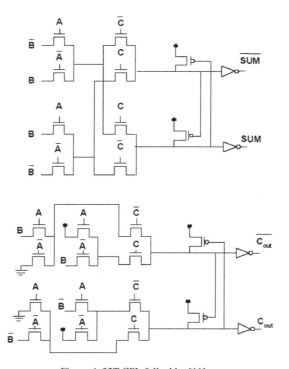

Figure 4. 32T CPL full adder [11].

88

CPL has been applied to the full adders in multiplier circuits and has been shown to result in high speed due to its low input capacitance and high logic functionality. However, when implementing CPL, particularly in reduced supply voltage designs, it is important to take into account the problems of noise margins and speed degradation. These are caused by mismatches between the input signal level and the logic threshold voltage of the CMOS inverters, which fluctuates with process variations. DPL is a modified version of CPL that meets the requirement of reduced supply voltage designs [16].

1.4. *The Double Pass transistor Logic (DPL) full adder*

The basic difference of pass transistor logic compared to the CMOS logic style is that the source side of the logic transistor networks is connected to some input signals instead of the power lines. In the Double Pass Transistor Logic (DPL) style [17-18], both NMOS and PMOS logic networks are used in parallel.
In Fig. 5, a DPL full adder is presented. It contains 48 transistors.

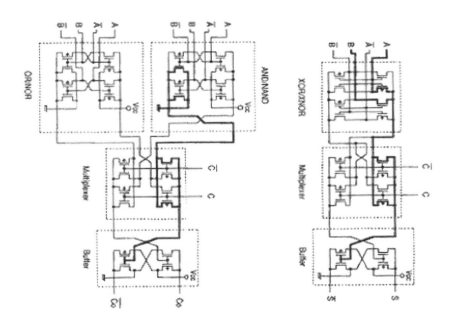

Figure 5. 48T DPL full adder [19].

1.5. *The full adder in combined technology (CMOS+ DPL)*

This proposed adder includes two sub circuits: The differential carry gate and the differential sum gate as shown in the target design presented in Fig. 7.

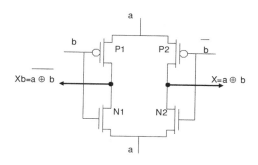

Figure 6. The differential XOR gate [20].

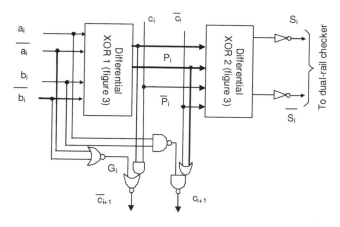

Figure 7. 28T full adder in combined technology (CMOS + DPL) [20].

This design combines the pass transistor CMOS technology with static CMOS technology.

The differential carry gate is designed in static CMOS with only 16 transistors [NIC 93]. As for the sum function $Si = ai \oplus bi \oplus ci$, it is implemented with two differential XOR. The first differential XOR performs signals $Pi = ai \oplus bi$ and $\overline{Pi} = ai \oplus bi$. These signals and the dual carry are the inputs of the second differential XOR that generates the dual sum function as shown in Fig.

This differential XOR is shown in Fig 6.

This fully differential implementation requires only 28 transistors.

2. The proposed self-checking full adder

In Fig. 8, the schematic of the proposed static DPL logic circuit for a full adder is shown.

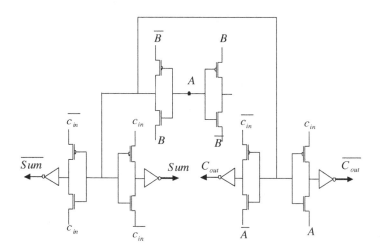

Figure 8. The proposed self-checking full adder [21].

Double pass-transistor logic is shown to improve circuit performance at reduced supply voltage. Its symmetrical arrangement and double-transmission characteristics improve the gate speed without increasing the input capacitance [19].

The full adder circuit is implemented in full-custom 32nm DPL technology [22]. SPICE simulations of the circuit extracted from the layout, including parasitic, are used to demonstrate that this adder has an acceptable and expected electrical behaviour.

The SPICE simulation of the differential full adder is as shown in Fig. 9.

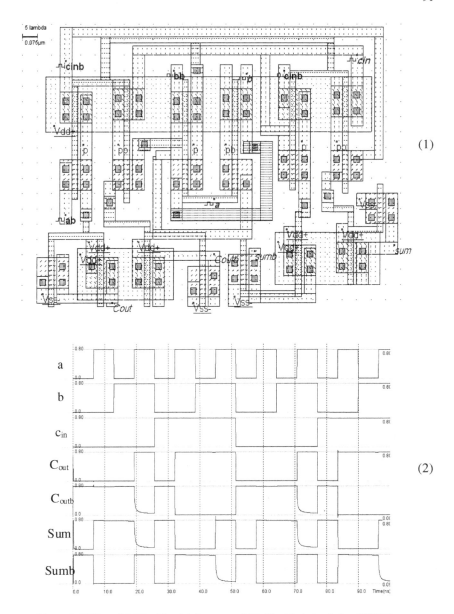

(1)

(2)

Figure 9. Differential full adder. (a): Layout and (b): Electrical simulation (SPICE).

The SPICE simulation of the differential full adder without inverters is as shown in Fig. 10.

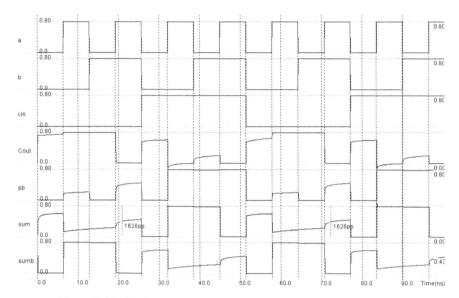

Figure 10. The SPICE simulation of the differential full adder without inverters.

To avoid this signal degradation, inverters are added in the outputs of the circuit.

3. Comparisons

In order to evaluate the usefulness of the proposed design and to demonstrate that this is an acceptable design style, we analyse in the following, the performance of the differential full adder shown in Fig. 8 in terms of number of transistor, area overhead and propagation delay.

3.1. Number of transistor

Table 1 summarizes the number of transistors for each type of adder.

Table 1. Number of transistors for each type of adder.

CMOS Full adder [11]	CMOS+DPL Full adder [20]	DPL Full adder [21]
60	28	20

The proposed design allows a decrease in the number of transistor. We save 66.66% of the transistors number overhead if we compare the proposed design to the duplication based adder.

3.2. *Area overhead*

Table 2 summarizes the overhead for each type of adder.

Table 2. Overhead for each type of adder.

	CMOS Full adder [11]	CMOS+DPL Full adder [20]	DPL Full adder [21]
Length (μm)	2.9	2,730	1.8
Width (μm)	1.7	1,770	1.4
Area (μm2)	4.8	4.83	2.5

The Layout of the CMOS full adder and the Layout full adder in combined technology [20] are shown respectively in Fig. 11 and Fig. 12.

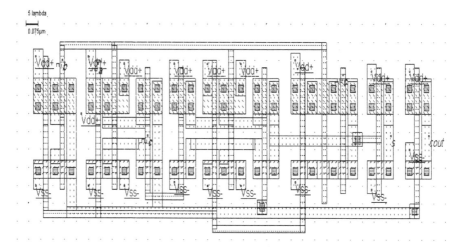

Figure 11. Layout of the CMOS full adder.

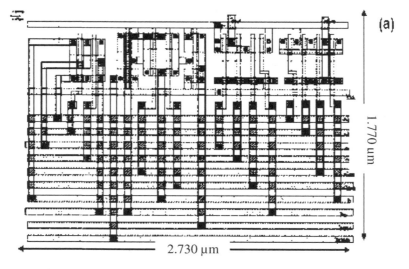

Figure 12. Layout of the full adder in combined technology (CMOS+ DPL) [20].

3.3. *Delay*

We analyse also, the response time of the circuit in term of rise and fall delays at the output node S_i. Fig. 13 gives rise and fall delays versus load capacitance.

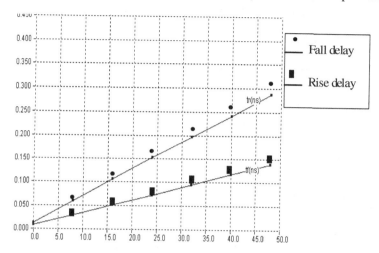

Figure 13. Delay vs. load capacitance.

References

1. M.V.Saideepika, S.Karthik and K.Priyadharshini, "Two phase clocked Adiabatic Logic for low power Multiplier," International Journal of Advanced Engineering Sciences and Technologies (IJAEST). Vol No. 5, Issue No. 2, 255 –260. (2011).
2. Jiang, Y., Al-Sheraidah, A., Yang, Y., Sha, E., and Chung, J.G., 'A Novel Multiplexer-based Low-power Full Adder', IEEE Transactions on Circuits and System II: Express Briefs, 51, 345–348. (2004).
3. Foroutan, V., Navi, K., and Haghparast, 'A New Low Power Dynamic Full Adder Cell Based on Majority Function', World Applied Sciences Journal, 4, 133-141. (2008).
4. Kowsalya, T., 'Tree Structured Arithmetic Circuit by using Different CMOS Logic Styles', ICGST International Journal on Programmable Devices, Circuits and Systems, 8, 11–18. (2008).
5. Ruholamini, M., Sahafi, A., Mehrabi, S., and Dadkhahi, N., 'Low-power and Highperformance 1-Bit CMOS Full-adder Cell', Journal of Computers, 3, 48–54. (2008).

6. Moaiyeri, M.H., Mirzaee, R.F., and Navi, K., 'Two New Low-power and High-performance Full Adders', Journal of Computers, 4, 119–126. (2009).

7. Navi, K., Saatchi, M.R., and Dael, O., 'A High-Speed Hybrid Full Adder', European Journal of Scientific Research, 26, 29–33. (2009).

8. Pooja Mendiratta and Garima Bakshi, "A Low-power Full-adder Cell based on Static CMOS Inverter," International Journal of Electronics Engineering, 2(1), pp. 143-149, (2010).

9. Chen Guo-zhang, Chen Hao and He Pi-lian,"Study and Evaluation in CMOS Full Adder," Transactions of Tianjin University. Vol. 9, NO. 1. March (2003).

10. Chang, C.H., J. Gu and M. Zhang, "A review of 0.18um full adder performances for tree structure arithmetic circuits". IEEE Trans. Very Large Scale Integration (VLSI) Syst., 13 (6): 686-695. (2005).

11. Keivan Navi and Omid Kavehei, "Low-Power and High-Performance 1-Bit CMOS Full-Adder Cell," Journal of Computers, vol. 3, no. 2, February. (2008).

12. Vahid Foroutan, Keivan Navi and Majid Haghparast, "A New Low Power Dynamic Full Adder Cell Based on Majority Function, " World Applied Sciences Journal 4 (1): 133-141, ISSN 1818-4952. pp. 133-141, (2008).

13. K. Yano, T. Yamanaka, T. Nishida, M. Saito, K. Shimohigashi, and A. Shimizu, "A 3.8-ns CMOS 16*16-b multiplier using complementary pass-transistor logic," IEEE. Custom Integrated Circuits Conference. (1989).

14. K. Yano, T. Yamanaka, T. Nishida, M. Saito, K. Shimohigashi, and A. Shimizu, "A 3.8-ns CMOS 16*16-b multiplier using complementary pass-transistor logic," IEEE J. Solid-State Circuits, vol. 25, pp. 388–395, Apr. (1990).

15. Vojin G. Oklobzija, "Differential and Pass-Transistor CMOS Logic for Hign-Performance Systems," Proc 21st International Conference on Microelectronis (MIEL'97). VOL. 2. Nis, Yugoslavia, 14-17 September (1997).

16. M. Suzuki, N. Ohkubo, T.Yamanaka, A. Shimizu, and K. Sasaki, "A 1.5ns 32b CMOS ALU in double pass-transistor logic, " in Proc. 1993 IEEE Int. Solid-State Circuits Conf. pp. 90-91. Feb. (1993).

17. N. Ohkubo et al., "A 4.4 ns CMOS 54*54-b multiplier using pass transistor multiplexer," IEEE J. Solid-State Circuits, vol. 30, pp. 251–257, Mar. 1995.

18. A. Bellaouar and M. I. Elmasry, "Low-Power Digital VLSI Design: Circuits and Systems", Kluwer, Norwell, MA, (1995).

19. M. Suzuki, N. Ohkubo, T.Yamanaka, A. Shimizu, and K. Sasaki, "A 1.5ns 32b CMOS ALU in double pass-transistor logic, " in Proc. 1993 IEEE Int. Solid-State Circuits Conf. pp. 90–91. , Feb. (1993).

20. Hamdi Belgacem, Khedhiri Chiraz, and Tourki Rached, "A novel differential XOR-based selfchecking adder," International Journal of Electronics, 2012, 1–23, iFirst, Taylor & Francis. (2012).

21. Chiraz Khedhiri, Mouna Karmani, Belgacem Hamdi, Ka Lok Man, Yue Yang and Lixin Cheng, "A Self-checking CMOS Full adder in Double Pass Transistor Logic," Proceedings of the International MultiConference of Engineers and Computer Scientists 2012 Vol II. IMECS 2012, 14-16 March 2012, Hong Kong.
22. E. Sicard, "Microwind and Dsch version 3.1," INSA Toulouse, ISBN 2-87649-050-1, Dec (2006).

USING THE WEB-CAMERA BASED EYE TRACKING TECHNOLOGY TO EXPLORE THE AUDIENCE'S ATTENTION PREFERENCES ON THE DIFFERENT LAYOUT COMPOSITIONS OF INFORMATION

HUI-HUI CHEN

Department of Computer and Communication Engineering, Ming Chuan University, Taoyuan, Taiwan, huichen@ mail.mcu.edu.tw

YI-TING YEH

Department of Computer and Communication Engineering, Ming Chuan University, Taoyuan, Taiwan, 00166021@ms1.mcu.edu.tw

CHIAO-WEN KAO

Department of Computer Science and Information Engineering, National Central University, Taoyuan, Taiwan, chiaowenk@gmail.com

BOR-JIUNN HWANG

Department of Computer and Communication Engineering, Ming Chuan University, Taoyuan, Taiwan, bjhwang@ mail.mcu.edu.tw

CHIN-PAN HUANG

Department of Computer and Communication Engineering, Ming Chuan University, Taoyuan, Taiwan, hcptw@mail.mcu.edu.tw

It could be more effectively to obtain human's inner complex cognitive process and visual information by recording and analyzing eye movements. This study used a web-camera based system, which is easy to obtain, unaware of its existence, and unnecessary to adjust the poses of the participants, to conduct the eye tracking experiment. While reading the different layout compositions of information, the audience's attention preferences of fixation position and duration, gaze areas, and navigation transitions were recorded and explored.

1. Introduction

1.1. *Perceiving Information of Users*

Along with the mature development of the digitalized information, there are plenty of learning materials and information offered via web pages. The integration of multimedia profoundly makes the presentation of information on the web in numerous ways.

No matter which kind of the integration takes place on the web, it only matters to the users whether the information could be delivered effectively to them [1]. Several research studies indicated that the users' eye gazes and navigation transitions were affected not only by the users' prior reading patterns and experiences but also affected by the arrangements of the page layouts[2], [3]. Hence, the arrangements of the page layouts influence the users' choices and distribution of their attentions caught by the layouts. It was found that the users read information by scanning. Only when they found interested regions, they would just further read closely [3]. They also spent more time gazing on the information with graphics and on the preferred regions of the layouts [4].

1.2. *Questionnaires Used for the Investigation on Eye Activities*

The study [5] indicated that during human's cognitive procedure on processing the information, there was more than 80 percent of the information acquired by eyes, which were the most important sources of the sensory memory. The studies on eye tracking to explore the relationship between the eye movements and the modalities to present information were published. Although the data collected from the eye movements were important, in the past the eye movements could only be observed through the eyes of the observers, or through the interviews and the think-aloud methods [6]. However, the past way not only might be easily misconducted by the false memory, but it also might not probably reflect a person's true inner cognition. Since a German psychologist, J. Cohn, published his investigation report on the roles of the favorite colors in the 19th century, most of the researchers inherited from him using questionnaires to investigate research variables about how and what eyes navigate [7].

However, the strategy of using questionnaires might cause the similar research studies with the opposite research results due to the different research methodologies, stimuli, or equipment. By using questionnaires in investigation, the results might also be influenced by the subjective impressions of people [8], [9].

1.3. *Eye Movement Tracking and Devices*

In the 20th century, researchers started to record eye movements from various aspects [6]. Recording a person's eye movement activities was recognized as a more objective and effective way to conduct research studies in how and what information eyes process [10].

Eye movements construct a series of visual tracks which reveal the paths that the eyesight of the audience navigates [2]. These visual tracks in turns may become the cues to understand the navigation patterns of the audience while reading. Recording and analyzing eye movements can effectively obtain human's inner complex cognitive process of visual information and can effective retrieve the foci of the processed information during reading [11], [12]. The final data can serve as the indicator of the outer behavior of the audience. Furthermore, through studying eye movements, that the interested gaze areas and to where the audience's attention pay may be acknowledged [5], [12], [13].

The eye tracking device, which early was used in studying reading in psychology, consists of two forms, the head mounted and the table mounted. The head mounted form is suitable for studying the motion related and the reality research of the 3D [6], but it is too heavy to be used in studying reading. Its weight may easily cause the neck of the participant uncomfortable [14] and cause the pressure and tiredness to the participant. Further, it will affect the participant's mood and make the data collected from recording eye movements inaccurate [15].

Therefore, the table mounted form is promoted generally for its ease to use and for that it affects the participant less while reading, although it can only record 2D tracks of the eye movement changes [6]. The disadvantage of the table mounted form is that it could only be placed on the table or the level area. While conducting research studies on reading with either the head mounted or the table mounted, it is necessary to adjust the visual angels or to limit the movement area of the participant. The participant cannot act at the participant's great convenience nor act as naturally as possible. Besides, both forms are addition to the purchase, not general enough seen. Therefore, to record eye movements, this research proposes to adopt the camera base form for its ease to obtain, its unawareness of existence, and the unnecessary for adjustment of the participant's poses [16]. The camera base form will be used to make the experiment closer to real condition to the participant and to construct a reliable and buildable experiment setting.

1.4. *Research Purpose*

To record eye movements, this research proposes to adopt the web-camera based form for its ease to obtain, its unawareness of existence, and the unnecessary for adjustment of the participant's poses. The web-camera based form are used to make the experiment closer to real condition to the participant and to construct a reliable and buildable experiment setting.

2. Research Design

2.1. *Research Framework*

This study will explore the audience's attention preferences (fixation position & fixation duration), gaze areas, gaze duration, and navigation transitions regarding to the layout compositions of photos and texts and their positions to the web pages (as shown in Figure 1).

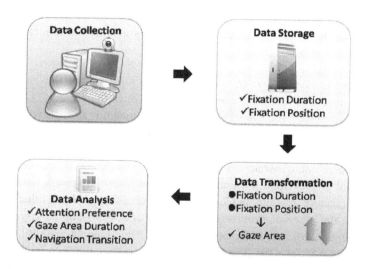

Figure 1. The 4 major works of the research.

1. Record the navigation transitions. The web camera will be initiated to record the eye movements while the participant starts to navigate the screen pages. Five shots will be taken per second since, based on the previous research studies, the average gaze time of reading the content in Chinese takes about 220-230 millisecond.

2. Record the fixation position & fixation duration for the audience's attention preferences.

3. Transform data (fixation position & fixation duration) into gaze areas.

4. Analyze the data. Explore the relationships between the layout compositions of photos and texts and their positions to the web pages in terms of the attention of preferences, the gaze area durations, and the navigation transitions.

2.2. *Research Questions and Method*

Regarding to the layout compositions of photos and texts and their positions to the web pages, this study comes up with three research questions, stated as followed:

- Research Question 1: What are the audience's attention preferences in terms of fixation position & fixation duration?
- Research Question 2: Whether the audience's attention is attracted to the possessed favorite type of the layout compositions of photos and texts?
- Research Question 3: Whether the audience's attention is attracted to the positions of information to the web pages?

Generally that in favor of eye tracking technology relies on eye movements can reflect the inner attention transformation process of the audience [11], [17]. It means that supervising the eye movements is corresponding to supervising when the cognitive process and attention initiate [18]. Despite of some research studies indicate that eye tracking records the fixation position, in fact human's visual attention could obtain information in a circumference area within 5-degree visual angel. Upon the common distance from the screen, this area is about 15 characters in width and 7 characters in height [19]. Some other studies indicate that the range of skimming information is about 4-degree visual angel while viewing photos and 2-degree visual angel while reading texts [6]. Hence, this study is to transform the fixation position and fixation duration into the gaze area and duration, and then to scratch the navigation transitions based on the gaze areas.

Psychologically proven, within a short period of time, one's favorite tendency will drive one's eye movements. One will gaze more time on one's more favorite colors and images, and one will repeat gazing on those [18]. In addition, to properly control variables which might affect the research results, this study surveys the most commonly types seen in web portal news of the layout compositions of photos and texts. Three layout compositions of photos and texts are presented in this study: the photo on the left with the texts on the right, the photo on the right with the texts on the left, and the photo on the top with the texts on the bottom. Moreover, to reduce personal subjective causes, news and photos, such as about idols, political parties, and etc., which might arouse the audience great interests, are eliminated. Taken place are news and photos about natural issues, general knowledge, and common leisure are chosen to be integrated into the designated content of the experiment layouts. Besides, the positions of the three layout compositions of photos and texts are equally shifted and arranged. Positions and layout compositions are crossly examined to answer the research questions. Hence, this study will extend to make the null hypotheses:

- Null Hypothesis H_01: There is no difference of attractions between the layout compositions of photos and texts in terms of the gaze area, gaze duration and navigation transition.
- Null Hypothesis H_02: There is no difference between the positions of information in terms of the gaze area, gaze duration and navigation transition.

This study adapted PVS (Pattern Voting Scheme) from a recent research study [22] to record eye tracking activities and collect data for PVS requires no special user action for looking at reference points. PVS is less complex in camera calibration process and cost less. PVS finds one's fixation on the screen by determining the spatial coordinates of the eyes by distinguishing between iris and sclera with pixel gray value of pixel despite of eye colors.

With PVS, first, the eye detection is carried out by using the method based on haar-like feature [23]. After finding the eye image, take the horizontal $\overline{AB}$ from the half height (as shown in Figure 2) and in order to make sure $\overline{AB}$ going through the iris position, $\overline{AB}$ position needs to be calibrated (as shown in Figure 3).

 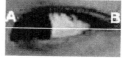

Figure 2. Shows (a) left eye (b) right eye take the horizontal $\overline{AB}$.

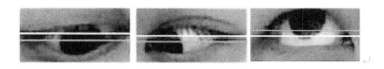

Figure 3. Calibrate $\overline{AB}$ to red line.

The screen then is divided into N*M areas, where N along the horizontal dimension and M along the vertical dimension. Let $\overline{AB}$ be divided into K*N equal segments, where K is an odd number that can be used to set the number of polling district. And then the mean value (MV_i) of gray value in each part can be obtained, where i from 1 to KN. Compute MV_i to obtain the voting weight to indicate the iris location and estimate the visual translation direction [23].

3. Experiments and Results

3.1. *Participants*

15 college students (7males, 8 females), age from 20 to 25, all are able to navigate the contents of web pages and basic reading ability [16].

3.2. *Experiment Materials*

A designed page consists of three layout compositions of photos and texts (as shown in Table 1):

- Composition A: the photo on the left with the texts on the right.
- Composition B: the photo on the right with the texts on the left.
- Composition C: the larger photo on the top with the texts on the bottom.

The webpage is divided into 3 blocks: Left (Block 1), Middle (Block 2), Right (Block 3). The positions of the 3 layout compositions of photos and texts are equally shifted and arranged respectively in the blocks. Therefore, there are 6 combinations (Com.1~Com.6) for the cross examination of the relationships between the positions of information and the layout compositions (as show in Table 2).

The contents of news and photos were retrieved from several web portals during November 25th to December 5th, 2011. News and photos which might arouse the audience subjective interests are eliminated to keep the experiment objective. In consideration of the major spoken language of the participants, the experimental page contents are in Chinese [16].

Table 1. Two samples of 3 types of layout compositions.

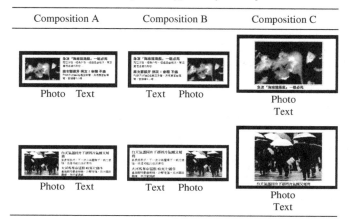

Composition A	Composition B	Composition C
Photo Text	Text Photo	Photo Text
Photo Text	Text Photo	Photo Text

Table 2. Combinations for the positions and layouts.

	Block 1 (Left)	Block 2 (Middle)	Block 3 (Right)
Com.1	A	B	C
Com.2	B	C	A
Com.3	C	A	B
Com.4	A	C	B
Com.5	B	A	C
Com.6	C	B	A

3.3. *Equipment*

One web camera, one personal computer, one LCD monitor.

3.4. *Experiment Procedure*

The experiment procedure is listed in the followings [16]:

1. Explain to the participants how to start and what to do.
2. When the participants finish navigating the presented information, they can click on the texts they gaze upon. Then, the next page will be shown with a waiting time of 2 seconds of blank page in black background to eliminate the effects caused by the previous residual image. That clicking on the texts

is imitated what generally the audience will do when navigating the news on the web.

3. Until they go through all of the six pages, the experiment ends.

3.5. *Data Analysis and Experiment Results*

The web camera took five shoots per second. Data were coded with the navigation time stamp. Total of valid 4616 shoots were collected. The data analysis and experiment results are explained as the following two sections of comparisons [16].

The first section is the comparisons for the respective positions of 3 layout compositions (A, B, C) on the webpage (as shown in Figure 4):

- Data showed that while A on the Block 1, eye gazes of the participants were taken the most times with 848 shoots.
- The gaze area of the participants mostly remained in Block 1, which is the left block on the webpage.
- There were no significant difference between A on the Block 2 and Block 3.
- The data revealed that the attention preferences of the participants did not follow where A went. Meanwhile similar situations happened to B and C as well.
- While B on the Block 1, eye gazes of the participants were taken the most times with 981 shoots.
- While C on the Block 1, eye gazes of the participants were taken the most times with 946 shoots.
- The data revealed that the attention preferences of the participants did not follow where B or C went, either.

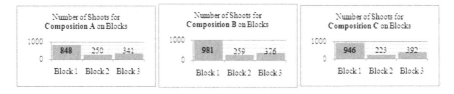

Figure 4. Comparisons for positions: Comparisons for the respective positions of 3 layout compositions (A, B, C) on the webpage.

The second section is the comparisons for the attention preferences of 3 layout compositions:

- For Composition A, B, C on the same block (as shown in Figure 5), the data showed that the eye gazes for Composition A, B, C on the same block received close attentions contemporarily. There were no significant differences among types of layout compositions. Whereas, the attention preferences turned to be attracted to whatever type of layout composition presented on Block 1, which is the left block on the webpage.

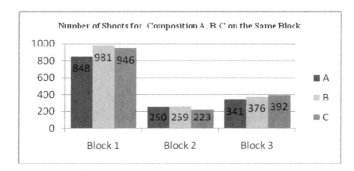

Figure 5. Comparisons for attention preferences via the eye gazes on the same block.

- For the cross comparisons for Composition A, B, C on different blocks (as shown in Figure 6), the data showed that the cross examined eye gazes for Composition A, B, C on the different blocks received close attentions contemporarily. There were no significant differences among types of layout composition.

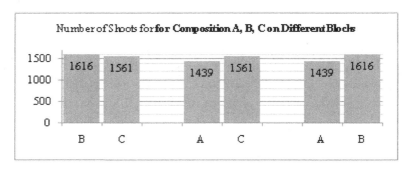

Figure 6. Cross comparisons for composition A, B, C.

- For the fixation position at start and the navigation transitions (as shown in Figure 7), the data showed that at the first 0.4 second, the fixation position at start mostly located on Block 1, which is the left block on the webpage.

Whereas, it could be noticed that even though during only 0.4 second at start, eyes navigated across the whole webpage and gazed much less on the middle block. About at the half time of navigation, the fixation position changed slightly.

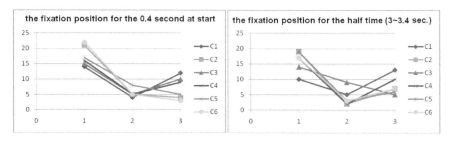

Figure 7. Fixation position at start and at the half time.

4. Conclusion

The gaze area of the participants mostly remained in the left block on the webpage. There were no significant difference of the gaze area between the middle block and right block. The result showed that the attention preferences of the participants were not affected by the layout compositions but affected by where the information was presented on the screen. In this case, the left block attracted the attention preferences of the participants. Therefore, information put on the left block of the screen can catch the audience's attention naturally.

Also, the result showed that the fixation position at start mostly located on the left block on the webpage. It also showed that eyes navigated across the whole webpage even at a glance of time but gazed much less on the middle block. Along with time, the fixation position changed slightly but data did not provide enough information to discover any pattern. In the future, it might be able to provide more insights if the experiment includes more participants.

References

1. R. C. Li, "A Study on Icon Semantics in Human-Computer Interaction," M. S. thesis, Dept. Industrial Design, Cheng Kung Univ., Taiwan (1993).
2. Mario T. Garica,, Contemporary newspaper design. Englewood Cliffs, NJ: Prentice Hall (1987).
3. K. Holmqvist, J. Holsanova, M. Barthelson, and D. Lundqvist, "Reading or Scanning A Study of Newspaper and Net Paper Reading," in The Mind's Eye: Cognitive and Applied Aspects of Eye Movement Research, R. Radach, J. Hyona, and H. Deubel, Eds. New York: Elsevier (2003).

4. N. Holmberg, "Eye movement patterns and newspaper design factors. An experimental approach," M. S. thesis, Lund University Cognitive Science (2004).

5. M. S. Sanders and E. J McCormick, Human Factors in Engineering and Design, New York: McGraw-Hill International Editions (1987).

6. H. C. Chen, H. D. Lai, and F. C. Chiu. "Eye Tracking Technology for Learning and Education," Journal of Research in Education Sciences, vol. 55, no. 4, pp. 39-68, Dec. (2010).

7. M. Perugini and R. Banse, "Personality, implicit self-concept and automaticity," European Journal of Personality, vol. 21, no. 3. pp. 257-261, Apr. (2007).

8. C. Taft, "Color meaning and context: comparisons of semantic ratings of colors on samples and objects," Color Research and Application, vol. 22, no. 1, pp. 40-50 (1996).

9. T. R. Lee, "New comparison of psychological meaning of colors in samples and objects with semantic ratings," Congress of the International Color Association, Rochester, NY, USA, Jun. 24 (2001).

10. Y. Zhu, Experimental Psychology, 2nd ed. Beijing: Peking University Press (2009).

11. J. M. Henderson and A. Hollingworth, "High-level scene perception," Annual Review of Psychology, vol. 50, pp. 243-271 (1999).

12. A. T. Duchowski, Eye tracking methodology: theory and practice, 2nd ed. New York: Springer (2007).

13. K. Rayner, "Eye movements in reading and information processing: 20 years of research," in Journal of Psychological Bulletin, vol. 124, no. 3, pp. 372-422 (1998).

14. The Advanced Knowledge Provider PITOECH CO.,LTD. (2009, Apr. 12). [On-line]. Available:
http://www.digitalwall.com/scripts/displaypr.asp?UID=13865

15. H. C. Chen, S. L. Peng, C. C. Tseng, and H. W. Chiou, "An Exploratory Study of the Relation Between the Average Saccade Amplitude and Creativity Under the Eyetracker Mechanism, " Bulletin of Educational Psychology, vol. 39, pp. 127-149 (2008).

16. H. H. Chen, Y.T. Yeh, C. W. Kao, B. J. Hwang, and C. P. Huang, "Through a web camera base of eye tracking technology to explore the audience's attention preferences in terms of the positions of information and the layout compositions," Lecture Notes in Engineering and Computer Science: Proceedings of The International MultiConference of Engineers and Computer Scientists 2012, IMECS 2012, 14-16 March, 2012, Hong Kong, pp. 224-228 (2012).

17. J. E. Hoffman, & B. Subramaniam," The role of visual attention in saccadic eye movements," Perception and Psychophysics, vol. 57, no. 6, pp. 787-795 (1995).

18. D. L. Tang, T. R. Lee, and C. M. Tsai. "An Exploratory Study on Relationship between Preference and Scanpath-Evidence from Color Preference Sorting Task," Chinese Journal of Psychology, vol. 47, no. 4, pp. 339-351, Dec. (2005).

19. B. Shneiderman and C. Plaisant, Designing the User Interface: Strategies for Effective Human-Computer Interaction, 4th ed. Boston: Addison-Wesley (2004).

20. 2010 Taiwan WEB 100 (2010, Apr. 02). [Online]. Available: http://163.26.9.13/trip/2010%E5%8F%B0%E7%81%A3web100%E6%8E %92%E5%90%8D%E5%88%86%E9%A1%9E.htm

21. Top 20 Websites for Different Ages in January, 2010 (2010, Mar. 03). [Online]. Available: http://blog.xuite.net/hinet_marketing/blog/32609798

22. Takeshi Mita, Toshimitsu Kaneko , Osamu Hori ," Joint Haar-like Features for Face Detection," Proc. of the Tenth IEEE International Conference on Computer Vision (2005).

23. C. W. Yang, C. W. Kao, K.C. Fan, B. J. Hwang, and C. P. Huang, "Eye gaze tracking based on pattern voting scheme for mobile device," presented at The First International Conference on Instrumentation & Measurement, Computer, Communication and Control, Beijing, China (2011).

HUMAN IDENTIFICATION BASED ON TENSOR REPRESENTATION OF THE GAIT MOTION CAPTURE DATA[*]

HENRYK JOSIŃSKI

Branch Faculty of Information Technology, Polish–Japanese Institute of Information Technology, Al. Legionów 2, Bytom, PL–41–902, Poland, hjosinski@pjwstk.edu.pl
Institute of Infomatics, Silesian University of Technology, Akademicka 16, Gliwice, PL–44–101, Poland, Henryk.Josinski@polsl.pl

ADAM ŚWITOŃSKI

Branch Faculty of Information Technology, Polish–Japanese Institute of Information Technology, Al. Legionów 2, Bytom, PL–41–902, Poland, aswitonski@pjwstk.edu.pl
Institute of Infomatics, Silesian University of Technology, Akademicka 16, Gliwice, PL–44–101, Poland, Adam.Switonski@polsl.pl

KAROL JĘDRASIAK

Branch Faculty of Information Technology, Polish–Japanese Institute of Information Technology, Al. Legionów 2, Bytom, PL–41–902, Poland, kjedrasiak@pjwstk.edu.pl

DANIEL KOSTRZEWA

Institute of Infomatics, Silesian University of Technology, Akademicka 16, Gliwice, PL–44–101, Poland, Daniel.Kostrzewa@polsl.pl

The authors present results of the research aiming at human identification based on gait motion capture data. Tensor objects were chosen as the appropriate representation of data. High-dimensional tensor samples were reduced by means of the multilinear principal component analysis (MPCA). For the purpose of classification the following methods from the WEKA library were used: k Nearest Neighbors (kNN), Naive Bayes, Multilayer Perceptron, and Radial Basis Function Network. The maximum value of the correct classification rate (CCR) was achieved for the classifier based on the multilayer perceptron.

1. Introduction

Gait is defined as coordinated, cyclic combination of movements which results in human locomotion [1]. A unique advantage of gait as a biometric is that it

---

[*] This work is supported by the research project OR00002111: "Application of video surveillance systems to person and behavior identification and threat detection, using biometrics and inference of 3D human model from video."

offers potential for recognition at a distance or at low resolution or when other biometrics might not be perceivable [2]. Gait can be captured by two-dimensional video cameras of surveillance systems or by much accurate motion capture[a] (*mocap*) systems which acquire motion data as a time sequence of poses.

Direct application of the mocap system for human identification is problematic because of the inconvenience of the capturing process. On the other hand, its great advantage is high precision of measurements. Thus, the usage of the mocap system in the development stage of the human identification system is reasonable [3].

Motion data lie in high-dimensional space [4], but the components of gait description, discussed in detail in section 2, are correlated, what allows dimensionality reduction.

The aforementioned problems formed the general objectives of the research: analysis of effectiveness of human identification based on gait mocap data with reduced dimensionality, and evaluation of the applied classification methods.

A full overview of bibliography describing the methods for solving the discussed problem would be unusually spacious. Generally, gait identification approaches can be divided into two categories: *model-free* and *model-based*. The former category can be split into approaches based on a moving shape and those which use integrate shape and motion within the description [2]. In the first example of the model-free approach silhouettes of walking human beings were extracted from individual frames using background subtraction, their morphological skeletons were computed and the modified independent component analysis (MICA) was proposed to project the original gait features from a high-dimensional measurement space to a lower-dimensional eigenspace. Subsequently, the L2 norm was used to measure the similarity between transformed gaits [5]. The principal components analysis (PCA) was also used in a similar way [6]. In [7] the recognition process was based on temporal correlation of silhouettes, whereas a spatio-temporal gait representation, called *gait energy image* (GEI), was proposed for individual recognition in [8]. The application of the *Procrustes* shape analysis method and the *Procrustes* distance measure in gait signature extraction and classification was shown in [9]. Numerous studies present frameworks developed for recognition of walking persons based on the dynamic time warping technique (DTW) [10], [11], as well as on the variants of the hidden Markov model (HMM), *inter alia*, generic HMM [12], population HMM [13], factorial and parallel HMMs [14].

---

[a] Motion capture is defined as "The creation of a 3D representation of a live performance" [27].

The model-based approaches use information about the gait, determined either by known structure or by modeling [2]. The ASF/AMC format is often applied as the skeleton model of the observed walking person. Numerous methods aim to estimate the model directly from two-dimensional images. In [15] the particle swarm optimization algorithm (PSO) is used to shift the particles toward more promising configurations of the human model. In [16] 2D motion sequences taken from different viewpoints are approximated by the Fourier expansion. Next, the PCA is used to construct the 3D linear model. Coefficients derived from projecting 2D Fourier representation onto the 3D model form a gait signature. Another set of features used for human identification is extracted from spatial trajectories of selected body points of a walking person (root of the skeleton, head, hands, and feet), named as *gait paths* [17].

It is stated in [18] that many classifiers perform poorly in high-dimensional spaces given a small number of training samples. Thus, feature extraction or dimensionality reduction is an attempt to transform a high-dimensional data into a low-dimensional equivalent representation while retaining most of the information regarding the underlying structure or the actual physical phenomenon [19]. The dimensionality reduction problem can be solved, *inter alia*, by encoding an image object as a general tensor of second or higher order [20]. The solution proposed in the aforementioned study includes the criterion for dimensionality reduction called *Discriminant Tensor Criterion* (DTC) and the algorithm called *Discriminant Analysis with Tensor Representation* (DATER).

Multilinear projection of tensor objects for the purpose of dimensionality reduction is the basis of the multilinear principal component analysis (MPCA). A survey with in-depth analysis and discussions is included in [21], whereas a framework for tensor object feature extraction is presented in [18]. One of the extensions of the MPCA – an unsupervised dimensionality reduction algorithm for tensorial data, named as *uncorrelated MPCA* (UMPCA) – is proposed in [22] and [23].

Tensor objects as a form of representation of gait sequences are discussed in section 2, whereas section 3 contains a brief description of the MPCA algorithm. Section 4 deals with procedure of the experimental research along with its results. The conclusions are formulated in section 5.

2. Tensor Representation of the Gait Mocap Data

Tensor object is a multidimensional object, the elements of which are to be addressed by indices. The number of indices determines the order of the tensor object, whereas each index defines one of the tensor modes. Gait silhouette sequences are naturally represented as third-order tensors with column, row, and time modes [18].

Description of each of the consecutive poses forming a gait sequence depends on the assumed skeleton model. For a typical model containing 22 segments and a global skeleton rotation (Fig. 1), description of a single pose comprises values of 69 Euler angles. Three additional values are required for specification of a global translation [3].

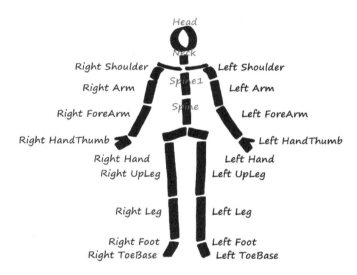

Figure 1. Components of the skeleton model.

The authors propose second-order representation of the gait motion capture data, composed of "time mode" and "pose mode". A single tensor object includes a single gait sequence built of 100 consecutive frames (poses) according to the requirement of the MPCA which accepts tensor samples of the same dimensions.

The global translation values were removed from the input data, what guarantees that the identification process is based solely on the body parts movement, not on the gait route. Additionally, values of the angles remaining constant for all consecutive poses were also eliminated as redundant.

Consequently, description of a single pose includes values of 51 Euler angles. Hence, total number of features characterizing a single gait sequence comes to 5100.

Gait sequences were recorded in the Human Motion Laboratory (HML) [24] of the Polish–Japanese Institute of Information Technology, equipped with the Vicon motion capture system (Fig. 2).

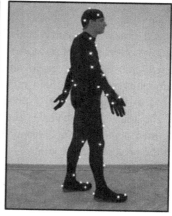

Figure 2. Mocap sessions in the Human Motion Laboratory.

3. The Multilinear PCA Algorithm

Multilinear projection of tensor objects for the purpose of dimensionality reduction was based on the algorithm and its MATLAB implementation presented in [18]. According to the authors of the MPCA algorithm: "Operating directly on the original tensorial data, the proposed MPCA is a multilinear algorithm performing dimensionality reduction in all tensor modes seeking those bases in each mode that allow projected tensors to capture most of the variation present in the original tensors" [18]. Its application leads to feature extraction by determining a *multilinear projection* – the mapping from a high-dimensional tensor space to a low-dimensional tensor space. A single point of a tensor object represents a single feature. Thus, number of features I in the N-order input tensor object is defined as $I = \prod_{n=1}^{N} I_n$, whereas after dimensionality reduction it is described by the formula $P = \prod_{n=1}^{N} P_n$, where $P_n \leq I_n$, $n \in [1, N]$. Symbols I_n, P_n denote the n-mode dimension of the tensor, respectively, before and after reduction.

As mentioned before, the tensor representation of the gait motion capture data proposed in this paper is composed of "time mode" (consecutive frames) and "pose mode" (significant Euler angles). Hence, as a result of the projection 2 *projection matrices* are constructed.

Computations were performed according to the MPCA algorithm in the following phases:

- Preprocessing – because all tensor samples are required to be of the same dimensions, an input set of 353 second-order tensor samples was normalized and, subsequently, centered by subtracting the mean value.
- Initialization (for each of 2 modes) – eigenvalues and eigenvectors were calculated. Subsequently, eigenvalues were arranged in descending order and cumulative sum of their relative contributions was computed and compared to the percentage Q of variation which should be kept in each mode. The first case, when the cumulative sum achieved or exceeded the user-defined value of Q, determined P_n (n = 1, 2) eigenvectors which formed the projection matrix.
- Local optimization of the projection matrices – improved versions of both projection matrices were computed one by one with the other one fixed.
- Projection of the centered input samples using the projection matrices – 353 *feature tensors* constituting the low-dimensional second-order representation of the input samples with Q % variation captured were obtained.

4. Experimental Research

Using the Vicon system 353 gait sequences for 25 men aged 20–35 years were recorded and stored in a database. The gait route was specified as a 5 meters long straight line. The acquiring process started and ended with a T-letter pose because of requirements of the Vicon calibration process. Two types of motion were distinguished: a slow gait and a fast one.

The mocap data were transformed into the second-order tensor representation. After the dimensionality reduction by means of the MPCA feature tensors were subject to the classification process.

The first purpose of the numerical experiments was to determine the dependency between the percentage Q of variation kept in each mode and the total number of features P resulting from the dimensionality reduction. The Q values were taken from the range of [80, 100] using a step value of 1 till $Q = 99$. The range [99, 100] was explored more deeply – a step value was set to 0.01. The obtained dependency between P and Q is presented for clarity in the logarithmic scale in Fig. 3.

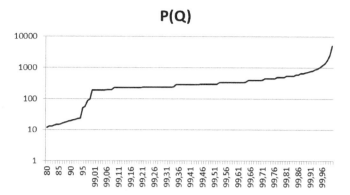

Figure 3. Dependency between P and Q in the logarithmic scale for the case of second-order tensors.

The gait motion capture data reduced by means of the MPCA were used in the first phase of the classification process by the following algorithms from the WEKA software [25]: 10 variants of the kNN ($k = 1..10$) and the Naive Bayes. The effectiveness of both methods expressed by means of the correct classification rate (CCR) was shown in Table 1 along with the most appropriate values of Q and P.

Table 1. Effectiveness of classification by means of the kNN and the Naïve Bayes method for the case of second-order tensors.

Classifier	CCR [%]	Q [%]	P
1NN	92.49	[99.21, 99.33]	240
2NN	90.88	[99.09, 99.20]	234
3NN	93.03	[99.35, 99.44]	287
4NN	92.22	[99.06, 99.08]	195
5NN	91.96	99.34	246
6NN	90.62	99.34	246
7NN	89.81	[99.06, 99.08]	195
8NN	88.74	[99.06, 99.08]	195
9NN	87.67	[99.00, 99.05]	190
10NN	86.06	[99.21, 99.33]	240
Naive Bayes	90.35	98.00	99

The dependency between CCR and the number k of neighbors taken into consideration by the kNN classifier was presented in Fig. 4.

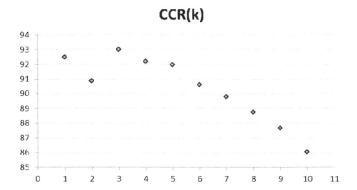

Figure 4. Correct classification rate for 10 kNN variants for the case of second-order tensors.

The effectiveness of the 3 best variants of the kNN classifier ($k = 1, 3, 4$) and of the Naive Bayes technique for the complete tested range of Q values was depicted in Fig. 5.

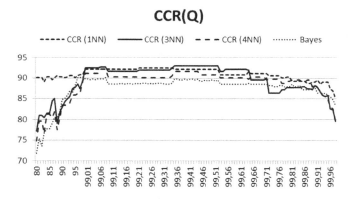

Figure 5. Dependency between CCR and Q for the most effective kNN variants and the Naive Bayes technique for the case of second-order tensors.

The best effectiveness (CCR = 93.03%) was obtained for the classifier 3NN using $P = 287$ features. As a consequence of this result, in the second phase of the classification process this value was assumed to be fixed. Thus, the methods applied in this phase – Multilayer Perceptron and Radial Basis Function Network – were used solely for classification of the feature tensors previously reduced to 287 features. Results of this process were shown in Table 2.

Table 2. Effectiveness of classification by means of the Multilayer Perceptron and the Radial Basis Function Network for the case of second-order tensors.

Classifier	CCR [%]	P
Multilayer Perceptron	95.71	287
Radial Basis Function Network	90.35	287

The maximum value of the CCR equal to 95.71% was achieved for the classifier based on the multilayer perceptron after 1600 epochs.

Inspired by the above-mentioned experimental results, analysed also in [26], the authors proposed the following assumptions for the next research stage:

- Third-order tensor representation with modes described, respectively, by numbers of components of Euler angles, numbers of skeleton components, and numbers of sequence frames.
- A single gait sequence was elongated to 128 frames.
- The Q values were taken solely from the range of [99, 100] using a step value of 0.01.
- 5 variants of the kNN ($k = 1..5$), the Naive Bayes and the Multilayer Perceptron methods were used for the purpose of classification.

The effectiveness of the 3 best variants of the kNN classifier ($k = 1, 2, 3$), the Naive Bayes (denoted by NB) and the Multilayer Perceptron (MLP) methods was depicted in Fig. 6.

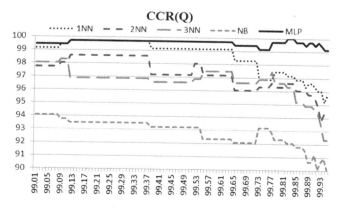

Figure 6. Dependency between CCR and Q for the most effective kNN variants, the Naive Bayes and the Multilayer Perceptron methods for the case of third-order tensors.

The best effectiveness (CCR = 100%) was obtained anew for the classifier based on the multilayer perceptron using the Q values from the range [99.82, 99.84]. However due to the long time of computations it is worthwhile to continue searching for more efficient classification methods.

5. Conclusion

In this paper the authors have discussed results of the research aiming at human identification based on gait motion capture data represented as second-order or third-order tensor objects. High-dimensional tensor samples were reduced by means of the MPCA and subsequently classified using kNN, Naive Bayes, Multilayer Perceptron, and Radial Basis Function Network. In this way the authors have tested an effective research procedure for gait analysis.

The sizeable reduction of dimensions of tensorial samples based on mocap data was achieved at the percentage of variation kept in each mode of only a little less than 100. Furthermore, classification based on the reduced number of features turned out to be more effective than at the full variation kept in each mode.

Future research will explore the influence of the feature selection methods on the effectiveness of the gait based identification process. Nonlinear techniques (Isomap, locally linear embedding (LLE)) are also planned to be applied for dimensionality reduction. Conclusions drawn from experiments with mocap data will be taken into account during the next stage of the research which will be carried out using video sequences.

References

1. J. E. Boyd and J. J. Little, "Biometric Gait Recognition", *Lecture Notes in Computer Science* 3161, Springer, (2005).
2. M. S. Nixon, T. N. Tan and R. Chellappa, *Human Identification Based on Gait*. Springer, (2006).
3. A. Świtoński, A. Polański, K. Wojciechowski, "Human Identification Based on the Reduced Kinematic Data of the Gait", *7th International Symposium on Image and Signal Processing and Analysis*, (2011).
4. J. Xiao, Y. Zhuang and F. Wu, "Getting Distinct Movements from Motion Capture Data", *Proceedings of the International Conference on Computer Animation and Social Agents*, (2006), pp. 33–42.
5. M. Pushpa Rani and G. Arumugam, "An Efficient Gait Recognition System for Human Identification Using Modified ICA", *International Journal of Computer Science & Information Technology*, Vol. 2, No. 1, (2010), pp. 55–67.

6. L. Wang, T. Tan, H. Ning and W. Hu, "Silhouette Analysis-Based Gait Recognition for Human Identification", *IEEE Transactions on Pattern Analysis and Machine Intelligence*, Vol. 25, No. 12, (2003), pp. 1505–1518.

7. S. Sarkar, P. J. Phillips, Z. Liu, I. R. Vega, P. Grother and K. W. Bowyer, "The HumanID Gait Challenge Problem: Data Sets, Performance, and Analysis", *IEEE Transactions on Pattern Analysis and Machine Intelligence*, Vol. 27, No. 12, (2005), pp. 162-177.

8. J. Han and B. Bhanu, "Individual Recognition Using Gait Energy Image", *IEEE Transactions on Pattern Analysis and Machine Intelligence*, Vol. 28, No. 2, (2006), pp. 316–322.

9. L. Wang, H. Ning, W. Hu and T. Tan, "Gait Recognition Based on Procrustes Shape Analysis", *The 9th International Conference on Image Processing*, (2002).

10. S. Hong, H. Lee, I. F. Nizami, S.-J. An and E. Kim, "Human Identification Based on Gait Analysis", *Proceedings of the International Conference on Control, Automation and Systems*, (2007), pp. 2234–2237.

11. A. Kale, N. Cuntoor, B. Yegnanarayana, A. N. Rajagopalan and R. Chellappa, "Gait Analysis for Human Identification", *Proceedings of the 4th International Conference on Audio- and Video-Based Biometric Person Authentication*, (2003), pp. 706–714.

12. A. Sundaresan, A. Roy–Chowdhury and R. Chellappa, "A Hidden Markov Model Based Framework for Recognition of Humans from Gait Sequences", *Proceedings of the 2003 IEEE International Conference on Image Processing*, (2003), pp. II–93–96.

13. Z. Liu and S. Sarkar, "Improved Gait Recognition by Gait Dynamics Normalization", *IEEE Transactions on Pattern Analysis and Machine Intelligence*, Vol. 28, No. 6, (2006), pp. 863–876.

14. Ch. Chen, J. Liang, H. Zhao, H. Hu and J. Tian, "Factorial HMM and Parallel HMM for Gait Recognition", *IEEE Transactions on Systems, Man, and Cybernetics – Part C: Applications and Reviews*, Vol. 39, No. 1, (2009), pp. 114–123.

15. T. Krzeszowski, B. Kwolek and K. Wojciechowski, "Articulated Body Motion Tracking by Combined Particle Swarm Optimization and Particle Filtering", *Lecture Notes in Computer Science* 6374, Springer, (2010), pp. 147–154.

16. Z. Zhang and N. F. Troje, "View-independent person identification from human gait", *Neurocomputing*, Vol. 69, (2005), pp. 250–256.

17. A. Świtoński, A. Polański and K. Wojciechowski, "Human Identification Based on Gait Paths", *Proceedings of the 13th International Conference on Advanced Concepts for Intelligent Vision Systems*, Springer, (2011).

18. H. Lu, K. N. Plataniotis and A. N. Venetsanopoulos, "MPCA: Multilinear Principal Component Analysis of Tensor Objects", *IEEE Transactions on Neural Networks*, Vol. 19, No. 1, (2008), pp. 18–39.

19. M. H. C. Law and A. K. Jain, "Incremental nonlinear dimensionality reduction by manifold learning", *IEEE Transactions on Pattern Analysis and Machine Intelligence*, Vol. 28, No. 3, (2006), pp. 377–391.

20. S. Yan, D. Xu, Q. Yang, L. Zhang, X. Tang and H.–J. Zhang, "Discriminant Analysis with Tensor Representation", *Proceedings of the 2005 IEEE Computer Society Conference on Computer Vision and Pattern Recognition*, (2005).

21. H. Lu, K. N. Plataniotis and A. N. Venetsanopoulos, "A Survey of Multilinear Subspace Learning for Tensor Data", *Pattern Recognition*, Vol. 44, No. 7, (2011), pp. 1540–1551.

22. H. Lu, K. N. Plataniotis and A. N. Venetsanopoulos, "Uncorrelated Multilinear Principal Component Analysis through Successive Variance Maximization", *Proceedings of the 25th International Conference on Machine Learning*, (2008).

23. H. Lu, K. N. Plataniotis and A. N. Venetsanopoulos, "Uncorrelated Multilinear Principal Component Analysis for Unsupervised Multilinear Subspace Learning", *IEEE Transactions on Neural Networks*, Vol. 20, No. 11, (2009), pp. 1820–1836.

24. *http://hm.pjwstk.edu.pl*: webpage of the Human Motion Laboratory of the Polish–Japanese Institute of Information Technology.

25. M. Hall, E. Frank, G. Holmes, B. Pfahringer, P. Reutemann and I. H. Witten, "The WEKA Data Mining Software: An Update", *SIGKDD Explorations*, Vol. 11, Issue 1, (2009).

26. H. Josiński, A. Świtoński, K. Jędrasiak and D. Kostrzewa, "Human Identification Based on Gait Motion Capture Data", Lecture Notes in Engineering and Computer Science: *Proceedings of The International MultiConference of Engineers and Computer Scientists 2012*, IMECS 2012, 14–16 March, 2012, Hong Kong, pp. 507–510.

27. A. Menache, *Understanding Motion Capture for Computer Animation and Video Games*. Morgan Kaufmann, (2000).

FORMAL MODELLING AND VERIFICATION OF COMPENSATING WEB TRANSACTIONS

SHIRSHENDU DAS, SHOUNAK CHAKRABORTY, HEMANGEE K. KAPOOR

Indian Institute of Technology Guwahati, Assam, India
* E-mail: {shirshendu, c.shounak, hemangee}@iitg.ernet.in

KA LOK MAN

Xi'an Jiaotong-Liverpool University, China and
Baltic Institute of Advanced Technologies, Lithuania
E-mail: ka.man@xjtlu.edu.cn

Correctness of a system depends on the correct execution of transactions. For large systems, e.g., web-based, these transactions involve more than one component and hence more than one independent entity. Such transactions are called Long Running Transaction(LRT) as they coordinate complex interactions among multiple sub-systems. Well known roll-back mechanism do not suffice to handle faults in LRTs, therefore compensation mechanisms are introduced. However, introduced structures are complex and hard to understand and handle. Formal methods are well known tool for modelling, analysis and synthesis of complex systems. In this chapter we present a technique that allows modelling LRTs using Compensating CSP, then translating them to Promela language and analyzing using SPIN tool. We have modelled and verified the LRTs. We exemplify it using a web service called Air-ticket Reservation System (ATRS).

Keywords: Formal model, Business transaction, Compensation, Process Algebra, cCSP, SPIN tool

1. Introduction

Business transactions typically involve coordination and interaction between multiple partners. These transactions need to deal with faults that can arise in any stage of the transactions. In usual database transactions, a roll-back mechanism is used to handle faults in order to provide atomicity to a transaction. However, for transactions that require long periods of time to complete, also called *Long Running Transactions (LRT)*, roll-back is not always possible. Handling faults where multiple partners involved are both difficult and critical. Due to their interactive nature, it is not possible to checkpoint LRTs, e.g. a sent message cannot be unsent. In such cases, a separate mechanism is required to handle faults. A possible solution of the problem would be that the system designer can provide a mechanism to compensate the actions that cannot be undone automatically.

Compensation is defined as an action taken to recover from error in business transactions or cope with a change of plan.[1] Compensations are installed for every committed activity in a long-running transaction. If one sub-transaction fails, then compensations of the committed sub-transactions in the sequence are executed in reverse order.

The coordination between business processes is particularly crucial as it includes the logic that makes a set of different software components become a whole system. Hence it is not surprising that these coordination models and languages have been the subject of thorough formal study, with the goal of precisely describing their semantics, proving their properties and deriving the development of correct and effective implementations.

Formal techniques proved their usefulness in quite a few areas, e.g. automotive industry,[2] electronics,[3,4] industrial devices control,[5] medical devices control.[6-8] Process calculi are models or languages for concurrent and distributed interactive systems. They have also been used for modelling interactions in latency insensitive SoC interconnects.[9] Being simple, abstract, and formally defined, process algebras make it easier to formally specify the message exchange between web services and to reason about the specified systems. Transactions and process calculi have met in recent years both for formalizing protocols as well as adding transaction features to process calculi.[10-13]

Fu et al.[14] propose a method that uses the SPIN model-checking tool. The SPIN[15] tool takes Promela (Process or Protocol Meta Language)[16] as the input language and verifies its LTL (Linear Temporal Logic)[17] properties. Interactions of the peers (participating individual web services) of a composite web service are modeled as conversations and LTL is used for expressing the properties of these conversations.

Several proposals have been made in recent years to give a formal definition to compensable processes by using process calculi. These proposals can be roughly divided into two categories. In one category, suitable process algebras are designed from scratch in the spirit of orchestration languages, e.g., BPEL4WS. Some of them can be found in.[18-20] In another category, process calculi like the π-calculus[21,22] and the join-calculus[23] are extended to describe the interaction patterns of the services where, each service declares the ways to be engaged in a larger process.

We discuss the technique for modelling business process in cCSP, then translating them into Promela language[16] and analyzing the model in SPIN tool.[15] A preliminary version of this method showing modelling of a car broker web service appears in.[24] In this chapter we introduce an informal translation from the cCSP to Promela, and leave formal description for the future research. We exemplify the process with a business web service called Air-Ticket Reservation System (ATRS).

In section 2 we discuss the syntax of cCSP with examples and model the ATRS in cCSP. Then we discuss Promela/SPIN in section 3. In this section we also discuss the mapping of cCSP model to Promela. In section 4 we discussed about the verification of different LTL properties on ATRS. We finalize the chapter with conclusions (section 5).

Standard Processes:

$$P, Q ::= \quad A \quad \text{(atomic event)}$$

|P; Q   (sequential composition)
|P□Q   (deterministic choice)
|P ⊓ Q   (non-deterministic choice)
|P||xQ   (parallel composition)
|SKIP   (normal termination)
|THROW   (throw an interrupt)
|YIELD   (yield to an interrupt)
|P ▷ Q   (interrupt handler)
|[PP]   (transaction block)

Compensable Processes:

$$PP, QQ ::= \quad P \div Q \quad \text{(Compensation Pair)}$$

|PP; QQ
|PP□QQ
|PP ⊓ QQ
|PP||xQQ
|SKIPP
|THROWW
|YIELDD

2. Compensating CSP (cCSP)

Transaction processing and process algebra inspired the development of process algebra cCSP.[20,25,26] cCSP is an extension of most popular process algebra called CSP,[27] which is used for modelling interactive systems. A subset of the original cCSP is considered in this chapter, which includes most of the operators, as summarized in the above table. Similar to CSP, processes in cCSP can engage in atomic events and can be composed using sequential; choice; and parallel composition operators. The processes are categorized into two types: (i) standard; and (ii) compensable: which have a separate set of actions to be executed upon failure of a transaction. Variables $P, Q, \ldots$ are used for standard processes and $PP, QQ, \ldots$ are used for compensable processes.

A simple process in cCSP consists of a set of atomic events. For example if ATM is a process having two events as $insertCard$ and $outputMoney$, then the behavior of ATM can be represented as:

$$ATM = insertCard \rightarrow outputMoney \rightarrow ATM$$

Here *insertCard* and *outputMoney* executes one after another and then the process (ATM) again goes to its initial state. To execute the process ATM, the user must initiate it by inserting his/her ATM card. Let the *User* process is:

$$User = insertCard \rightarrow getMoney \rightarrow SKIP$$

Here $SKIP$ means successful termination of a process. The above two processes have one common event called *insertCard*. Now to withdraw money both *User* and ATM must simultaneously agree with the *insertCard* event. i.e., ATM must ready to accept the card when *User* inserts it. In cCSP this behavior is defined as:

$$SYSTEM = ATM \|_{insertCard} USER$$

This is called parallel composition in cCSP. If two process P and Q are in parallel composition $(P\|_x Q)$ then they must simultaneously agree with the occurrence of their common events (x). Events in cCSP can also assume as communication channels, through which two process will communicate. In the above process $(SYSTEM)$, the two subprocesses are composed with a parallel operator and they have a common event $(insertCard)$. So communication through the channel *insertCard* is only possible if ATM is ready to accept the information set by the $USER$.

In a process P, input on channel a and output on channel b can be described as $P?a$ and $Q!b$ respectively. $P; Q$ is a process composed of two processes P and Q. This is called sequential composition. Here execution of Q starts only after the successful termination of P. Note that P and Q must be processes and not events. For example $a; Q$ is not a sequential composition, because a is not a process. The valid composition should be $a \rightarrow SKIP; Q$. But for shortcut we use the first representation $(a; Q)$ in this chapter.

In the case of failures in long running transactions, we need support to raise interrupt and handle the interrupt. The *THROW* action is used to raise an interrupt and *YIELD* is used to handle it. For example, $(P; YIELD; Q)$ is willing to yield to an interrupt in between the execution of P and Q.

A compensable process is constructed using a pair $(P \div Q)$, where P is the forward behavior used to model normal execution, and Q is the associated compensation designed to compensate actions executed in P. The sequential composition is defined in such a way, that actions done in P are accumulated and will be executed in reverse order in case composition needs to be aborted and compensated. By enclosing a compensable process PP inside a transaction block $[PP]$, we get a complete transaction, where the transaction block is also a standard process. Successful completion of PP represents successful completion of the block. But, when the forward behavior of PP throws an interrupt, the compensations are executed inside the block, and the interrupt is not observable from outside the block.

2.1. *Air-ticket Reservation System (ATRS)*

In this section we have taken ATRS as the web-service to model using cCSP. Our ATRS is inspired from[28] but not exactly the same. We have changed/removed/added some functionality in ATRS. The flow diagram of ATRS is given in figure 1. It has four services: **Traveler, Agent, Airline** and **Bank**. Traveler initiates the execution by expressing his interest of booking an air ticket. It requests Agent to book ticket. Agent on receiving the order, requests Airline for seat availability query. If seats available then Airline replies to Agent, who will immediately forward information to the Traveler. If no seats are available or Traveler suddenly decides for not booking, then Traveler raises an exception. Otherwise, Traveler will send a request to the Agent. Agent will confirm the Airline regarding booking. After getting confirmation, sends the payment request to the Agent. Agent will add its charge with the payment and send it to the Traveler. Traveler will send his account credentials to the Agent which will be sent to the bank by the Agent. Bank will check the credentials, and if it is valid then it will send success to the Agent, else Bank throws an exception. After getting success from the Bank, Agent will send the payment request. Bank will check the payment amount and if it satisfies the condition(s) then Bank pays without throwing an exception. Bank informs both Agent and Traveler after payment procedure is over. Agent, then informs Airline regarding successful payment and gets ticket from Airline. Agent forwards the ticket to the Traveler. If Traveler wants to cancel the booking then he will send a cancel request to the Airline via Agent. Airline will cancel the ticket if terms and conditions are not violated and send refund information to the Agent. Agent forwards the refund information to the Bank. Bank, after refunding, sends a refund message to Traveler. In each case whenever any service raises an exception, the normal execution of all the services will stop and corresponding compensation actions start in each service model.

Figure 1 shows each request and response channels in chronological order. For clarity each channel's name and chronological order number are intersecting the channel in at-least one point. While modelling we have assumed the whole ATRS system under one transactional block. Hence, exception raised by any process will be automatically yield by other processes.

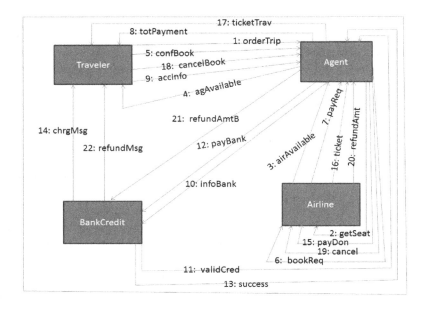

Fig. 1.   Processes and connecting channels in ATRS

The cCSP model of ATRS also consist of four processes, e.g., **Traveler**, **Agent**, **Airline** and **Bank**. We already discussed the functionality of these modules. Now we present each of these processes one by one.

2.1.1. cCSP Model for Traveler

The process **Traveler** starts execution by sending a message through the channel ordertrip and waits for a reply through agAvailable. Note that all the messages sent from **Traveler** are accepted by **Agent**. In other words, all outgoing channels from **Traveler** are connected to **Agent** as input. Similarly all messages received by **Traveler** are sent from **Agent** (except chrgMsg). After getting information through agAvailable it either confirms the booking or declines the booking. For declining the booking it throws an exception and the normal flow of the process suspends immediately and the corresponding compensation process **CompensateTraveler** starts. On the other hand, if **Traveler** confirms the booking then it sends confirmation message through confBook and wait for reply via totPayment. After getting the reply, **Traveler** will send account information through accInfo and waits for the money-receipt and ticket through chrdMsg and `ticketTrav` respectively. After getting the ticket, **Traveler** can non-deterministically accept or cancel the ticket. To cancel the ticket it sends a cancel message through `cancelBook` and waits for the money refund message through `refundMsg`. The process will terminate successfully if it accepts the ticket or receives refund message via `refundMsg`. Note that whenever **Traveler** sends message to **Agent**, it interacts with other two processes (**Airline** and **Bank**), as required, before sending reply to the **Traveler**.

$$
\begin{aligned}
\textbf{Traveler} \quad &\cong \quad \text{orderTrip!;} \\
&\qquad ((\text{agAvailable?1;confBook!}) \\
&\qquad \square \\
&\qquad (\text{agAvailable?0; THROWW})); \\
&\qquad \text{totPayment?}t; \text{accInfo!}i; \text{chrgMsg?}; \text{ticketTrav?}; \\
&\qquad ((\text{cancelBook!; refundMsg?}) \sqcap \text{SKIPP})) \\
&\qquad \div \textbf{CompensateTraveller} \\
\textbf{CompensateTraveller} \quad &\cong \quad \text{SKIPP}
\end{aligned}
$$

2.1.2. cCSP Model for Agent

The process **Agent** communicate with other three processes to co-ordinate the whole booking procedure. It starts executing, once it receives a booking request from Traveler through `orderTrip`. **Agent** will throw an exception if payment through **Bank** is not successful.

$$
\begin{aligned}
\textbf{Agent} \quad &\cong \quad \text{orderTrip?; getSeat!;} \\
&\qquad ((\text{airAvailable?1; agAvailable!1}) \\
&\qquad \square \\
&\qquad (\text{airAvailable?0; agAvailable!0})); \\
&\qquad \text{confBook?; bookReq!; payReq?}r; \text{totPayment!}t; \\
&\qquad \text{accInfo?}i; \text{infoBank!}i; \text{validCred?; payBank!}t; \\
&\qquad ((\text{success?1; payDon!; ticket?; ticketTrav!}) \\
&\qquad \square \\
&\qquad (\text{success?0; THROWW})); \\
&\qquad ((\text{cancelBook?; cancel!1; refundAmt?}a; \text{refundAmtB!}b; \text{SKIPP}) \\
&\qquad \square \\
&\qquad (\text{SKIPP})) \\
&\qquad \div \textbf{CompensateAgent} \\
\textbf{CompensateAgent} \quad &\cong \quad \text{SKIPP}
\end{aligned}
$$

2.1.3. cCSP Model for Airline

This process starts its execution once it receive seat availability query request from **Agent** through `getSeat`. After getting payment confirmation it non-deterministically either sends the ticket or throws an exception.

Airline	$\cong$	getSeat?; airAvailable!k; bookReq?; payReq!s;
		payDon?; (SKIPP $\sqcap$ THROWW); ticket!;
		((cancel?1; **refundProcess**;)
		$\square$
		(cancel?0; SKIPP))
		$\div$ **CompensateAirline**
CompensateTraveller	$\cong$	SKIPP
refundProcess	$\cong$	refundAmt!a ; SKIPP

2.1.4. *cCSP Model for Bank*

This process starts its execution once **Agent** forwards traveler's banking information via `infoBank`.

Bank	$\cong$	(infoBank?i; (validCred! $\sqcap$ THROWW); payBank?t;
		(success!1 $\sqcap$ success!0);)
		$\square$
		(refundAmtB?a; refundMsg!)
		$\div$ **CompensateBank**
CompensateBank	$\cong$	SKIPP

3. Modeling in Promela

Promela is the modelling language used in the Spin tool. It is used to model the required interaction behavior and verify properties. The model consists of processes and channels. Processes are independent entities which need to be invoked using the `run` clause. Processes interact with each other over message channels and/or globally declared variables. Variables can be of types: `bit`, `bool`, `byte`, `array` etc. For details of all data types see the Promela manual in.[15,16]

The behavior of a process is defined by a *proctype* declaration and instantiated using the `run` command.

```
proctype A() { byte state; state = 3;}
init
{ run A(); }
```

The keyword `atomic` makes all the enclosed statements to be executed as one indivisible unit, non-interleaved with any other processes.

```
atomic{statements;}
```

Message channels are either input or output and carry data between processes. For example, the channel `myout` outputs value of variable `a`, whereas the channel `myinput` reads the incoming value in the variable `b`.

```
chan myout = [2] of byte;
chan myinput = [0] of byte;
myout!a ; myinput?b ;
```

Table 1: Translation from cCsp to Promela

Constructs of cCSP	Promela
Process This is a cCSP process.	`run Process();`
Process(param) This is a cCSP process with parameters.	`run Process(param);`

P ÷ Comp
To implement this in Promela we assume a global variable *comp-cond*. When *comp-cond* will be true execution of *P* will be suspended and *Comp* will start.
	`proctype P{`
	`//statements of P.`
	`}unless{`
	`(comp_cond==true)`
	`run Comp();`
	`}`

P ∥ Q Each process in Promela runs in parallel if not mentioned otherwise.	`run P();` `run Q();`
P ; Q We used two *atomic* statements for representing sequential composition.	`atomic{run P();}` `atomic{run Q();}`

event1 □ event2 or event1 ⊓ event2
These are not cCSP processes. They are deterministic/non-deterministic choice between two events of a cCSP process.
	`if`
	`:: event1;`
	`:: event2;`
	`fi;`

input-channel?value
This is also just an event of a process. In this event *input-channel* waits for an input and stores it into variable *value* when it gets the input.
	`int value;`
	`// Other data types possible.`
	`input_channel?value;`

output-channel!value
In this event *output-channel* is ready to send the data stored in value. It can only send when the corresponding input channel is also ready to receive it.
	`int value;`
	`// Other data types possible.`
	`output_channel!value;`

THROW To implement THROW from a process we set a global variable *comp-cond* as true. Once the *comp-cond* becomes true the first statement after unless will be true and compensation process will start executing. See implementation of *P ÷ Q* above in this table.	

The channel capacity can be given after its name. In the above example, channel `myout` has buffer capacity of 2 and the channel `myinput` having capacity zero is used for rendezvous communication. As cCSP uses rendezvous communication, we have used similar channels in our model.

For control flow, the `if` statement does a selection between a set of options. If multiple options are enabled, then any one is chosen at random. If none of the options are enabled, then the statement blocks until some statement becomes executable. In the following example, any one option will get executed, depending on the condition.

```
if
 :: (a != b) -> option1;
 :: (a == b) -> option2;
fi
```

The most important statement of Promela that we used in this chapter is the *unless* statement

```
{ statements1 } unless { statements2 }
```

It starts execution in statements1. Before every statement in statements1 is executed, it checks if the first statement in statements2 can be executed. If yes, then the control transfers to statements2 else it continues execution of statements1. If statements1 terminates, statements2 is ignored.

3.1. *Translation from cCSP to Promela*

In this section we discuss briefly about how to model a cCSP process to Promela. Table 1 shows all the important mappings from cCSP to Promela.

3.2. *Promela model for ATRS*

Due to space constraints we show the Promela model for only the **Traveler** process.

```
proctype Traveller()
{
  bit agav=0,cnfb=0,cb=0,rfm=0,chm=0,nocan=0;
  { orderTrip!1; agAvailable?agav; // Send booking order and wait.
    if // if seat available then non-deterministically
       // accept or reject booking.
     :: (agav==1) ->
         if
             :: cnfb=1; // cnfb=1 means accept.
             :: cnfb=0; // cnfb=0 means reject.
         fi;
     :: (agav==0) -> travellerCompensate=1; //Raise exception if no seats.
    fi;
    if // booking process continues if cnfb=1, otherwise raise exception.
     :: (cnfb==1) -> confBook!cnfb;
     :: (cnfb==0) -> travellerCompensate=1;
    fi;
    totPayment?t; accInfo!i; chrgMsg?chm; ticketTrav?ticket;
    if
     :: (chm==1 && ticket==1) -> travellerBookSuccess=1; printf("Tkt Booked");
         if
             :: printf("Not canceling ticket."); nocan=1;
             :: cancelBook!1; refundMsg?rfm;
           if
             :: (rfm==1) -> travellerCancelSuccess=1;
             :: (rfm==0) -> travellerCompensate=1;
           fi;
       fi;
     :: (chm==0 || ticket==0) -> travellerCompensate=1;
    fi;
    if
     :: (airlineSuccess==1 && agentSuccess==1 && bankSuccess==1) ->
         if
```

```
          :: ((travellerBookSuccess==1 && nocan==1)||(travellerBookSuccess==1
             && travellerCancelSuccess==1)) -> travellerSuccess==1;
       fi;
    fi;
} unless{
  (travellerCompensate==1 || agentCompensate==1 || bankCompensate==1 ||
   airlineCompensate==1)-> printf("Traveler compensation started.");
}
```

The first statement of the **unless** part in the above code checks the condition for compensation. As soon as the condition is true it starts executing the compensation actions of Traveler. We have not modeled the compensation actions, we just printed a message that "compensation started". Note that the compensation condition depends on the exception raised by all the four processes (Traveler, Agent, Bank and Airline).

4. Verification

We have verified safety and liveness properties for the model. The verification output for liveness is given below. Liveness is checked by showing the absence of acceptance cycles and also non-progress cycles. The model was free of acceptance cycles:

```
(Spin Version 6.2.2 -- 6 June 2012)
        + Compression
Full statespace search for:
        never claim             - (not selected)
        assertion violations    +
        acceptance   cycles     + (fairness disabled)
        invalid end states      +
State-vector 200 byte, depth reached 70, errors: 0
    1768 states, stored
    3431 states, matched
    5199 transitions (= stored+matched)
       3 atomic steps
hash conflicts:        0 (resolved)
```

4.1. *Property Verification*

The list of LTL properties satisfied by the model is given below:

(1) When Traveler process will compensate, eventually all others processes will also be canceled.

```
ltl ltl_0 { [] ((travellerCompensate==1) -> (<> ( (travellerCancel==1) &&
            (agentCancel==1) && (airlineCancel==1) && (bankCancel==1))))}
```

Verification output for this property is given below.

```
(Spin Version 6.2.2 -- 6 June 2012)
        + Partial Order Reduction
Full statespace search for:
        never claim                + (ltl_0)
        assertion violations       + (if within scope of claim)
        acceptance   cycles        - (not selected)
        invalid end states         - (disabled by never claim)
State-vector 208 byte, depth reached 124, errors: 0
    1659 states, stored
    1673 states, matched
    3332 transitions (= stored+matched)
       3 atomic steps
hash conflicts:        0 (resolved)
```

(2) It will always happen that, when Agent process will compensate, eventually all others processes will also be canceled.

```
ltl ltl_1 { [] ((agentCompensate==1) -> (<> ((travellerCancel==1) &&
        (agentCancel==1) && (airlineCancel==1) && (bankCancel==1))))}
```

(3) With the compensation of Airline process, eventually all others processes will also be canceled.

```
ltl ltl_2 { [] ((airlineCompensate==1) -> (<> ((travellerCancel==1) &&
        (agentCancel==1) && (airlineCancel==1) && (bankCancel==1))))}
```

(4) Compensation of Bank process eventually implies the cancelation of all others processes.

```
ltl ltl_3 { [] ((bankCompensate==1) -> (<> ((travellerCancel==1) &&
        (agentCancel==1) && (airlineCancel==1) && (bankCancel==1))))}
```

(5) Success of Traveler will always imply, success of all other processes, eventually.

```
ltl ltl_4 { [] ((travellerSuccess==1) -> (<> ((agentSuccess==1) &&
        (airlineSuccess==1) && (bankSuccess==1))))}
```

Verification output for this property is given below.

```
(Spin Version 6.2.2 -- 6 June 2012)
        + Partial Order Reduction
Full statespace search for:
        never claim                + (ltl_4)
        assertion violations       + (if within scope of claim)
        acceptance   cycles        - (not selected)
        invalid end states         - (disabled by never claim)
State-vector 208 byte, depth reached 124, errors: 0
    1208 states, stored
     696 states, matched
    1904 transitions (= stored+matched)
       3 atomic steps
hash conflicts:        0 (resolved)
```

5. Conclusion

Modelling, analysis and implementation of complex business transactions is not a trivial task. In this chapter we present a technique that could be helpful in solving this problem. We propose to use cCSP for modelling of business transactions, then to translate cCSP model to Promela and to analyze it using SPIN. In such a way a language designed for such processes can be used for modelling (cCSP) and then, a well known and mature tool (SPIN) can be used for analysis of the system. We have defined a procedure for translating cCSP model to the Promela language and exemplified using a realistic Air-ticket Reservation System (ATRS) example. We also verified different properties for the implemented Promela model of ATRS.

References

1. J. Gray and A. Reuter, *Transaction Processing : Concepts and Techniques.* Morgan Kaufmann Publishers, 1993.
2. B. Gebremichael, T. T. Krilavičius, and Y. S. Usenko, "A formal model of a car periphery supervision system in UPPAAL," in *Proc. of Workshop on Discrete Event Systems*, 2004, pp. 433–438.
3. H. K. Kapoor, "Formal modelling and verification of an asynchronous dlx pipeline," in *The 4th Int, Conf. on Software Engineering and Formal Methods (SEFM)*, 2006, pp. 118–127.
4. K. Man, T. Krilavičius, C. Chen, and H. Leung, "Application of bhave toolset for systems control and mixed-signal design," in *Proc. of the Int. MultiConference of Engineers and Computer Scientists (IMECS)*, Hongkong, March 2010.
5. T. Krilavičius and V. Miliukas, "Functional modelling and analysis of a distributed truck lifting system," in *The 5th Int. Conf. on Electrical and Control Technologies (ECT 2010)*, Kaunas, Lithuania, 2010, p. 6.
6. T. Krilavičius, D. Vitkute-Adžgauskienė, and K. Šidlauskas, "Simulation of the radiation therapy system for respiratory movement compensation," in *Proc. of the 7th Int. Conf. Mechatronic Systems and Materials (MSM 2011)*, Kaunas, Lithuania, July 2011.
7. T. Krilavičius and K. Man, "Timed model of the radiation therapy system with respiratory motion compensation," in *The 6th Int, Conf. on Electrical and Control Technologies (ECT 2011)*, Lith., 2011, p. 6.
8. K. Man, T. Krilavičius, K. Wan, D. Hughes, and K. Lee, "Modeling and analysis of radiation therapy system with respiratory compensation using Uppaal," in *Proc. of the 9th IEEE Int. Symp. on Parallel and Distributed Processing with Application (ISPA 2011)*, Korea, 2011.
9. H. K. Kapoor, "Process algebraic view of latency-insensitive systems," *IEEE Transactions on Computers*, vol. 58, no. 7, pp. 931–944, 2009.
10. M. Berger and K. Honda, "The two-phase commitment protocol in an extended pi-calculus." *Electr. Notes Theor. Comput. Sci.*, 2000.

11. A. Black, V. Cremet, R. Guerraoui, and M. Odersky, "An equational theory for transactions," in *In Proc of FSTTCS*. Springer, 2003, pp. 38–49.

12. L. Bocchi, C. Laneve, and G. Zavattaro, "A calculus for long-running transactions," in *Formal Methods for Open Object-Based Distributed Systems*, ser. LNCS, E. Najm, U. Nestmann, and P. Stevens, Eds. Springer, 2003, vol. 2884, pp. 124–138.

13. R. Bruni, C. Laneve, and U. Montanari, "Orchestrating transactions in join calculus," 2002.

14. X. Fu, T. Bultan, and J. Su, "Analysis of interacting bpel web services," in *Proc. of the 13th Int. Conf. on World Wide Web*, 2004, pp. 621–630.

15. Basic spin manual. [Online]. Available: http://spinroot.com/spin/Man/Manual.html

16. Promela manual. [Online]. Available: http://spinroot.com/spin/Man/promela.html

17. B. Banieqbal, H. Barringer, and A. Pnueli, "Temporal logic in specification," in *LNCS*. UK: Springer, 1989, pp. 8–10.

18. R. Bruni, H. Melgratti, and U. Montanari, "Theoretical foundations for compensations in flow composition languages," *SIGPLAN Not.*, vol. 40, pp. 209–220, January 2005.

19. M. Butler and C. Ferreira, "A process compensation language," in *Integrated Formal Methods*, ser. LNCS, W. Grieskamp, T. Santen, and B. Stoddart, Eds. Springer Berlin / Heidelberg, 2000, vol. 1945, pp. 61–76.

20. M. Butler, C. Hoare, and C. Ferreira, "A trace semantics for long-running transactions," in *25 Years of CSP*, A. Abdallah, C. Jones, and J. Sanders, Eds., vol. Lectur. Springer, October 2005, pp. 133–150.

21. R. Milner, J. Parrow, and D. Walker, "A calculus of mobile processes, i," *Inf. Comput.*, vol. 100, pp. 1–40, September 1992.

22. J. Parrow, *Handbook of Process Algebra*. Elsevier, 2001, ch. 8, pp. 479–543.

23. C. Fournet and G. Gonthier, "The reflexive cham and the join-calculus," in *In Proc. of the 23rd ACM Symp. on Principles of Programming Languages*. ACM Press, 1996, pp. 372–385.

24. K. Wan, H. K. Kapoor, S. Das, B. Raju, T. Krilavicius, and K. L. Man, "Modelling and verification of compensating transactions using the spin tool," in *Proc. of The International MultiConference of Engineers and Computer Scientists (IMECS)*, Hong Kong, March 2012.

25. S. Ripon, "Extending and relating semantic models of compensating csp," Ph.D. dissertation, University of Southampton, UK, August 2008. [Online]. Available: http://eprints.ecs.soton.ac.uk/16584/

26. S. H. Ripon and M. Butler, "Formalizing ccsp synchronous semantics in pvs," *Society*, p. 9, 2010. [Online]. Available: http://arxiv.org/abs/1001.3464

27. C. A. R. Hoare, "Communicating Sequential Processes," in *Prentice-Hall International Series in Computer Science*, 1998.

28. G. Diaz, J.-J. Pardo, M.-E. Cambronero, V. Valero, and F. Cuartero, "Verification of web services with timed automata," *Electron. Notes Theor. Comput. Sci.*, vol. 157, no. 2, pp. 19–34, May 2006.

A MACHINE LEARNING APPROACH FOR CLASSIFICATION OF INTERNET WEB SITES[*]

AJAY S. PATIL

School of Computer Science, North Maharashtra University
Jalgaon, Maharashtra 425 001, India, aspatil@nmu.ac.in

B.V. PAWAR

School of Computer Science, North Maharashtra University
Jalgaon, Maharashtra 425 001, India, bvpawar@nmu.ac.in

Subject based web directories classify web pages into various hierarchical categories. This classification is done with the help of several human editors. Since important pages are properly classified according to subject categories, it is popular among the web user community. But the exponential growth of the web is making it difficult to manage subject based web directories with human editors. Although the World Wide Web (WWW) has a Domain Name System (DNS), it lacks a comprehensive directory of web sites. Web site classification using machine learning techniques is therefore an emerging possibility to automatically maintain directory services for the web. Home pages of web sites are distinguished pages which act as entry points to the rest of the pages of the web site. Information contained in title, keyword, description, anchor (A HREF) tags along with other content is a rich source of features required for classification. The web site developer puts in more effort on the design of the home page as compared to the rest of the pages of the web site. The designing of the home page includes giving the web site an aesthetic look, proper links to secondary pages and provide precise summary of the organization to which the site belongs. These special characteristics of web site home page can be exploited to identify the type of the organization to which the web site belongs. Humans easily deduce the organization type by viewing the contents of the home page of a web site. In this paper we present an experimental study to classify web sites based on the content of their home pages.

1. Introduction

Since its inception in 1991, the WWW has grown rapidly and is becoming popular amongst the information providers and users. Netcraft report estimates the web to have 584 million web sites registered (175 million active) in the beginning of 2012. The world Internet usage growth has increased by 480.4% during 2000-2011 (Internet World Stats). Thousands of search tools are

---

[*] This work is supported by University Grants Commission, New Delhi, India under research project scheme for teachers.

available on the Web to locate information. These tools can be classified into three major types based on their design viz., 1. Crawler based Search Engines (SE) e.g., Google, Bing, Yahoo etc., 2. Meta Search Engines e.g., Metacrawler, Clusty etc., and 3. Subject Directories like DMOZ (Directory Mozilla), Librarians Internet Index (LII) etc. The architectural features of these tools are detailed in Table 1 given below.

Table 1. Architectural features of searching tools

Features	Crawler based search engine	Meta search engines	Directories
Coverage	Indexable web	Indices of selected search engines	Web pages added according to predefined categories
Data Collection Mechanism	Crawler-Indexer	Forward-aggregate	Mostly human assisted
Resource Type	Text and multimedia	Text and multimedia	Mostly text
Data refresh	During next refresh cycle of the crawler	Search engine dependent	Poor
Data Repository	Huge repository of web documents	No repository of own	Relatively very small
Ranking	According to query relevance	Affected by ranking strategies of all the participating SEs	According to categories
Search	Keyword based	Keyword based	Browsing
Results	Too many	Many	Very few
Quality of Results	User needs time to find relevant ones	User needs time to find relevant ones	Good quality results, due to accurate Classification

Although subject directories are popular, manual classification is expensive to scale and is highly labor intensive. DMOZ [1] (world's largest subject directory) has 93,431 editors for one million categories and has indexed 4.98 million websites, which is less than three percent of the total active web sites. In order to make such directories scalable it is necessary to automate the process of maintaining subject directories. Search tools fail to answer queries like listing organizations related to a particular business or situated in a particular region. However if websites are classified (web site directory) according to the different categories it would be possible to answer such queries. The existence of a Web site directory can contribute to improve the quality of search results, help in filtering of web content, development of knowledge bases, building efficient focused crawlers or vertical (domain specific) search engines.

This paper describes Naïve Bayesian (NB) machine learning approach for the automatic classification of web sites based on content of home pages. The

NB approach, is one of the most effective and straightforward methods for text document classification and has exhibited good results in previous studies conducted for data mining.

2. Related Work

In the early days, classification was done manually by domain experts. But very soon, classification was also carried out in semi-automatic or automatic manner. Some of the approaches for text-categorization include statistical and machine leaning techniques like k-Nearest Neighbor approach [2], Bayesian probabilistic models [3]-[4], inductive rule learning [5], decision trees [4],[6], neural networks [7],[8] and support vector machines [9],[10]. While most of the learning methods have been applied to pure text documents, there are numerous publications dealing with classification of web pages. Pierre [11] discusses various practical issues in automated categorization of web sites. Machine and statistical learning algorithms have also been applied for classification of web pages [12]-[15]. In order to exploit the hypertext based organization of the web page several techniques like building implicit links [16], removal of noisy hyperlinks[17], fusion of heterogeneous data[18], link and context analysis[19] and web summarization[20] are used. An effort has been made to classify web content based on hierarchical structure [21].

3. Classification of Web Pages

Classification of web content is different in some aspects as compared with text classification. The uncontrolled nature of web content presents additional challenges as compared with classification of traditional text. Web content is semi structured and it contains several tags for formatting and providing hyperlinks to other web documents. The web documents undergo various cleaning steps (include removal of HTML tags, punctuation marks and stop-words) followed by stemming to make them ready for classification [26]. Machine learning algorithms interpret these documents as frequency vectors and use them to train the respective classifier. The classification component tests an unlabelled sample document against the learnt data. Our approach focuses on using home pages of organizational websites for learning and classification. A neatly developed home page of a web site is treated as an entry point for the entire web site. It represents the summary of rest of the web site. Many URLs link to the second level pages telling more about the nature of the organization. The title, keyword, description and anchor tags contain features rich in information. Most of the homepages are designed in a manner to fit in a single viewable screen. Such features contribute to the expression power of the home page to identify the nature of the organization.

4. Bayes Theorem

Consider $D=\{d_1,d_2,d_3,...d_p\}$ to be a set of documents and $C=\{c_1,c_2,c_3,....c_q\}$ be set of classes. Each of the p number of documents in D are classified into one of the q number classes from set C. The probability of a document d being in class c using Bayes theorem is given by:

$$c_{map} = \arg\max_{c \in C} P(c \mid d) = \arg\max_{c \in C} \frac{P(c)P(d \mid c)}{P(d)}$$

As $P(d)$ is independent of the class, it can be ignored.

$$= \arg\max_{c \in C} P(c)P(d \mid c)$$

Assuming that the attributes (terms) are independent of each other,

$$P(d \mid c) = P(t_1 \mid c)P(t_2 \mid c)P(t_3 \mid c)...P(t_{n_d} \mid c)$$
$$= \prod_{1 \le k \le n_d} P(t_k \mid c)$$

$$c_{map} = \arg\max_{c \in C} P(c \mid d)$$
$$= \arg\max_{c \in C} P(c)P(d \mid c)$$
$$= \arg\max_{c \in C} P(c) \prod_{1 \le k \le n_d} P(t_k \mid c)$$

$P(C)$ prior probability of c is N_c/N. Where N_c is training documents in class c and N is the number of training documents. $P(c \mid d)$ is posterior probability of c, as it reflects our confidence that c holds after we have seen d.

$$P(t \mid c) = \frac{T_{ct}}{\sum_{t' \in V} T_{ct'}}$$

Here, T_{ct} is the number of occurrences of t in D from class c, and $\sum_{t' \in V} T_{ct'}$ is the total number of terms in D from class c.

A term-class combination that does not occur in the training data makes the entire result zero. In order to solve this problem we use add-one smoothing or *Laplace smoothing*. The equation after adding Laplace's correction becomes:

$$P(t \mid c) = \frac{T_{ct} + 1}{\sum_{t' \in V}(T_{ct'} + 1)} = \frac{T_{ct} + 1}{\sum_{t' \in V}(T_{ct'}) + |V|}$$

Also, in order to prevent floating point underflows, summation of logs of probabilities is used instead of multiplying probabilities. Class with highest final un-normalized log probability score is the most probable.

5. Experimental Setup

The entire machine learning experiment explained below begins with a collection of website homepages classified into different categories. We prepare the required dataset after cleaning these homepages by eliminating the html tags, scripts, style sheets etc. This dataset is later subjected to training and testing of classifier. These steps are briefly discussed as under.

5.1. *Creation of Data Set*

Our data set consists of home pages in HTML (Hypertext Markup Language) format belonging to 10 different categories mentioned in Table 2. In order to create the dataset, services of various search engines and subject directories were used. The popular search engines like Google, Bing, Altavista etc., were submitted keyword based queries and then the results obtained were examined. If any of the links in the results pointed to a homepage, we visually examined its contents and if it belonged to the category of our interest, it was saved in respective directories. Since our classification is purely text based, home pages of some web sites that made hug use of Flash graphics or made used of other plug in applications were avoided. Home pages in languages other than English and having size less than 200 characters were also ignored. Hereafter we shall refer to these home pages as documents for simplicity. The entire dataset was then independently subjected to two annotators for moderation. Documents were removed from the dataset wherever there was a disagreement between the annotators. The dataset consisted of 4887 documents in ten different categories.

Table 2. Data Set

Category	Total Samples	Training Samples	Test Samples
Academic Institutions	503	453	50
Hotels	470	423	47
Book Sellers/Publisher	485	437	48
Health Care	511	460	51
Sports	476	428	48
Automobiles	495	445	50
Tours & Travel	475	427	48
Computer	502	452	50
Banking	490	441	49
Domestic Appliances	480	432	48
Total	4887	4398	489

5.2. *Cleaning HTML Documents*

The Jericho HTML Parser (Version 3.2) [23] was used to extract the HREF (hyperlink) label, TITLE, META DESCRIPTION and META KEYWORD and all BODY text containing in each document of the dataset. The Jericho HTML Parser is an open source library released under both the Eclipse Public License (EPL) and GNU Lesser General Public License (LGPL). This library is available online on the internet at http://jerichohtml.sourceforge.net. The Jericho HTML Parser 3.2 is a powerful java library allowing analysis and manipulation of parts of an HTML document, including server-side tags. One advantage of using this library is that the presence of badly formatted HTML does not interfere with the parsing of the rest of the document. In many web site images were as buttons to be clicked in place of hyperlinks or images were used to display name of the organization. Such information is a very important feature for classification purpose, however our experiment concentrates on text based retrieval so such graphical text we ignored.

The standard stop word list used in Bow [23] was used. Bow is a library of C code useful for writing statistical text analysis, language modeling and information retrieval programs. Words in plural format were converted into their singular version using similar approach. We applied stemming by constructing a map of words and their relevant stems. We did not use stemming algorithm such as Porter as in most cases stemming totally changes the meaning of the word and in some cases it is undesired. e.g., In case of the book category the word "book" refers to a physical book, whereas in the hotel category the word "booking" which also stems to "book", refers to "reservation".

6. Vocabulary Generation

Common features that are part of every web site were considered as stop features (About Us, Home, All Rights Reserved, Contact Us, Feedback etc). Such words are similar to regular stop words but specific to home pages. Some of these home page specific stop words are mentioned in table 3.

Table 3. Home page specific stop words

Information, login, view, browser, website, web, online, search, keyword, designed, copyright, rights, reserved, click, search, welcome, email, click, contact, developed, mail, home, page, feedback, webmaster …

Such words were also considered as stop words and therefore were removed from the dataset. It was also observed that webmasters inflated the title, meta description and keyword tags with multiple keywords. We normalized such repeating keywords to reduce the impact of site promotion techniques applied by webmasters. This step was performed during the cleaning phase. A vector called as vocabulary containing the most relevant words (features) was created for the experiment. The relevant words or features were those words that occurred more than seven times in the entire training set. Thus very rare words and also the very common words (stop words) were eliminated. The vocabulary count was 4500 for 4,398 documents of the training set. We term this vocabulary set as V. The next section discusses training and testing of the classifier.

6.1. *Training the Classifier*

The K fold strategy (with $k=10$) was followed to decide the number of training and testing examples. Nine folds i.e., 4398 examples were used as the training set to build the classifier and the remaining fold 489 examples were used to test the classifier for accuracy. The prior probability for each category is $1/10$ (as there are 10 categories). The posterior probability $P(wk|c)$ was calculated as follows. All documents that belonged to respective categories were parsed and a hash table was prepared for each category. All words in the vocabulary served as keys of the hash table. The values of the hash table were the word occurrence frequency (nk) in all documents belonging to that category. The total word count (including repeats) for each category termed as n was also calculated. The posterior probability with Laplace's correction was calculated using the formula $P(wk|c) = (nk + 1) / (n + |Vocabulary|)$. Partial feature sets generated as an effect of the experiment for academic and sports categories are given in Table 4 and 5.

Table 4. Sample feature set for academic institutions category

university, school, department, syllabus, student, alumina, placement, examination, result, principal, chancellor, campus, registrar, library, study, course, information, education, PG, center, technology, conference, administration, workshop, science, commerce, faculty, programme, academic,....

Table 5. Sample feature set for sports category

cricket, sports, score, goal, stadium, ground, kit, ball, umpire, referee, stumps, hockey, football, badminton, wrestling, player, commentary, highlight, victory, win, won, team, wicket, field, game, match, penalty, corner, kick, service, court, seed, scorecard, tour, champion,...

6.2. *Testing the Classifier*

In order to classify a document say X, the probabilities of a given category are looked up in the hash table and multiplied together. The category producing the highest probability is the classification for document X. Only the words found in X would be looked up in the hash table. Also if a word in X is absent in the original vocabulary (built from training set) the word is ignored. The equation used to classify X is $C = arg\ max\ (\ P(c)\ \Pi\ P(wk|c)\ )$. The Naïve Bayes algorithms to train and test the classifier are as given below:

ALGORITHM NB TRAINING
___

1. Let V be the vocabulary of ALL words in the documents in D

2. For each category $c_i\ \in\ C$
     Let Di be the subset of documents in D in category c_i
     $P(c_i) = |D_i|\ /\ |D|$
     Let T_i be the concatenation of all the documents in D_i
     Let n_i be the total number of word occurrences in T_i
     For each word $w_j \in\ V$
          Let n_{ij} be the number of occurrences of w_j in T_i
          Let $P(w_j\ |\ c_i) = (n_{ij} + 1)\ /\ (n_i + |V|)$

ALGORITHM NB TESTING
___

1. Given a test document X
2. Let n be the number of word occurrences in X
3. Return the category:
$$arg\ max\ P(c_i)\ =\ \prod_{i=1}^{n} P(a_i\ |\ c_i)$$
$$c_i \in C$$
     where a_i is the word occurring at the ith position in X

7. Experimental Results

Table 6 shows the results obtained when nine folds i.e., 4398 examples were used as the training set to build the classifier and the remaining fold 489 examples were used to test the classifier for accuracy. We use recall [24], precision [24] and F-measure [24] to verify the accuracy of our classification approach. F-measure is the harmonic mean of recall and precision. Recall, Precision and F-Measure are calculated as follows:

$$Recall = \frac{Number\ of\ relevant\ documents\ retrieved}{Total\ number\ of\ relevant\ documents}$$

$$Precision = \frac{Number\ of\ relevant\ documents\ retrieved}{Total\ number\ of\ documents\ retrieved}$$

$$\text{F-measure} = \frac{(2 \times \text{Recall} \times \text{Precision})}{(\text{Recall} + \text{Precision})}$$

Using the three measures, we observe that the average precision is 89.09%, average recall is 89.04%, whereas the F-Measure is 89.05%. Thus, classification of web sites is possible by examining the contents of their home pages.

Table 6. Classification accuracy

Category	Precision	Recall	F-Measure
Academic Institutions	93.36	89.46	91.37
Hotels	89.29	90.43	89.86
Book Sellers	88.66	88.66	88.66
Hospitals	85.13	85.13	85.13
Sports	93.36	88.66	90.95
Automobiles	88.65	89.90	89.27
Tours & Travel	90.21	89.26	89.73
Computer Dealers	86.44	87.65	87.04
Banks	88.89	89.80	89.34
Domestic Appliances	86.93	91.46	89.14
Average	89.09	89.04	89.05

7.1. *Training Examples and Accuracy*

The classifier was subjected to training and testing in 9 steps each time increasing the input by 50 documents. Figure 1 depicts number of training examples versus the accuracy in terms of average F-measure.

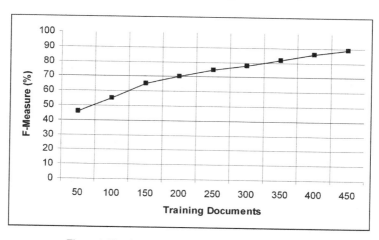

Figure 1. Number of training documents versus accuracy

The accuracy of the classifier was very poor (45%), when only 50 documents were supplied as training data. The accuracy increased each time when the classifier was supplied with additional learning data. The classifier achieved an accuracy of 89% when nearly 450 documents were supplied as input in each category. Thus, the accuracy of the classifier depends on the number of training documents and in order to achieve high accuracy, the classifier should be supplied with sufficiently large training documents.

8. Conclusion and Future Work

NB approach used in this paper exploits the richness of features of a home page of a website for classification into industry type category. It categorizes the web pages into very broad categories. NB approach for classification of home pages for the ten categories considered above yielded 89.05% accuracy. The classification accuracy of the classifier is found proportional to number of training documents. This approach can be used by search engines for effective categorization of websites to build an automated website directory based on type of organization. However we considered only distinct and non hierarchical categories. The same technique could also be used to classify the pages into more specific categories (hierarchical classification) by changing the feature sets.

References

1. DMOZ open directory project. [Online]. Available: http://dmoz.org/
2. G. Guo, H. Wang and K. Greer, "An kNN model-based approach and its application in text categorization", *5th Int. Conf., CICLing* Springer, Seoul, Korea, 2004, pp. 559-570.
3. McCallum and K. Nigam, "A comparison of event models for Naïve Bayes text classification", in *AAAI/ICML-98 Workshop on Learning for Text Categorization*, 1998, pp. 41-48.
4. D.D. Lewis and M. Ringuette, "A Classification of two learning algorithms for text categorization", in *Proc. of 3rd Annual Symposium on Document Analysis and Information Retrieval (SDAIR'94)*, 1994, pp. 81-93.
5. S.T. Dumais, J. Platt, D. Heckerman, and M. Sahami, "Inductive learning algorithms and representations for text categorization", in *Proc. of the 17th Int. Conf. on Information and Knowledge Management (CIKM'98)*, 1998, pp. 148-155.
6. C. Apte and F. Damerau and S. M. Weiss, "Automated learning of decision rules for text categorization", *ACM Trans. on Information Systems*, Vol. 12, no.3, pp. 233-251, 1994.
7. S. Wermter, "Neural network agents for learning semantic text classification", *Information Retrieval*, Vol. 3, no. 2, pp. 87 - 103, Jul 2000.
8. A.S. Weigend, E.D. Weiner, and J.O. Peterson, "Exploiting hierarchy in text categorization", *Information Retrieval*, Vol. 1, no. 3, pp.193-216, 1999.
9. E. Leopold, and J. Kindermann, "Text categorization with support vector machines. How to represent texts in input space?", *Machine Learning,* Vol. 46, no. 1-3, pp. 423-444, 2002

10. D. Bennett and A. Demiritz, "Semi-Supervised support vector machines", *Advances in Neural Information Processing Systems*, Vol. 11, pp. 368-374, 1998.
11. J. M. Pierre, "Practical issues for automated categorization of web sites.", in *Electronic Proc. of ECDL 2000 workshop on the Semantic Web*, Lisbon, Portugal, 2000.
12. Sun, E. Lim and W. Ng, "Web classification using support vector machine", in *Proc. of the 4th Int. workshop on Web information and data management*, McLean, Virginia, USA, 2002, pp. 96 – 99.
13. Y. Zhang and B. F. L. Xiao, "Web page classification based on a least square support vector machine with latent semantic analysis", in *Proc. of the 5th Int. Conf. on Fuzzy Systems and Knowledge Discovery 2008*, Vol. 2, pp. 528-532,
14. O. Kwon and J. Lee, "Web page classification based on k-nearest neighbor approach", in *Proc. of the 5th Int. Workshop on Information Retrieval with Asian languages*, Hong Kong, China, 2000, pp. 9-15.
15. S. Dehghan and A. M. Rahmani, "A classifier-CMAC neural network model for web mining", in *Proc. of the IEEE/WIC/ACM International Conference on Web Intelligence and Intelligent Agent Technology 2008*, Vol. 1, pp. 427-431.
16. S. Dou, S. Jian-Tao , Y. Qiang and C. Zheng, "A comparison of implicit and explicit links for web page classification", in *Proc. of the 15th International Conference on World Wide Web*, Edinburgh, Scotland, 2006 , pp. 643–650.
17. S. Zhongzhi and L. Xiaoli, "Innovating web page classification through reducing noise", *Journal of Computer Science and Technology*, Vol. 17, no. 1, pp. 9–17, Jan. 2002
18. Z. Xu, I. King and M. R. Lyu, "Web page classification with heterogeneous data fusion", in *Proc. of the 16th International Conference on World Wide Web*, Banff, Alberta, Canada, 2007, pp. 1171 – 1172,
19. G. Attardi, A. Gulli, and F. Sebastiani, "Automatic web page categorization by link and context analysis", in *Chris Hutchison and Gaetano Lanzarone (eds.), Proc. of THAI'99, 1999*, pp. 105-119.
20. S. Dou, C. Zheng, Y. Qiang, Z. Hua-Jun, Z. Benyu, L. Yuchang and M. Wei-Ying, "Web-page classification through summarization", in *Proc. of the 27th annual Int. ACM SIGIR Conf. on Research and Development in Information Retrieval*, Sheffield, United Kingdom, 2004, pp. 242 - 249.
21. S. Dumais and H. Chen, "Hierarchical classification of web content", in *Proc. of the 23rd annual Int. ACM SIGIR Conf. on Research and Development in Information Retrieval*, Athens, Greece, 2000, pp. 256 - 263.
22. The Jericho HTML Parser Library Version 3.2, [Online]. Available: http://www.jerichohtml.sourceforge.net
23. The BOW or libbow C Library [Online]. Available: http://www.cs.cmu.edu/~mccallum/bow/
24. C. J. van Rijsbergen. (1979). Information Retrieval (2nd ed.) London: Butterworths, [Online]. Available: http://www.dcs.gla.ac.uk/Keith/Preface.html
25. T. M Mitchell. (1997). Machine Learning McGraw-Hill Companies, Inc.
26. S. Patil, B. V. Pawar,"Analysis of traditional information retrieval techniques applied to the world wide web", *International Journal of Computer Science and Information Technology*, Vol. 1 no. 2, pp. 63-73, 2008
27. S. Patil, B. V. Pawar, Lecture Notes in Engineering and Computer Science: *Proceedings of The International MultiConference of Engineers and Computer Scientists 2012*, IMECS 2012, 14-16 March, 2012, Hong Kong, pp 519-523

WEB SERVICES FOR CHRONIC PAIN MONITORING

NUNO GONÇALO COELHO COSTA POMBO

Department of Informatics,
University of Beira Interior, Covilhã, Portugal
E-mail: ngpombo@ubi.pt
www.ubi.pt

PEDRO JOSÉ GUERRA DE ARAÚJO

IT-Institute of Telecommunications, Department of Informatics,
University of Beira Interior, Covilhã, Portugal
E-mail: paraujo@di.ubi.pt

JOAQUIM MANUEL VIEIRA DA SILVA VIANA

Faculty of Health Sciences
University of Beira Interior, Covilhã, Portugal
E-mail: jsviana@fcsaude.ubi.pt

The use of web services allows, anywhere and at anytime, a truly global, platform independent, and interoperable mean to access information. This chapter presents an overview of the key concepts for electronic pain diaries, the role of web services and its integration in the computerized system to monitorize chronic pain patients. The usage of web services may lead to enhance therapeutic assertiveness, through improving the process of acquisition and sending data, as well as the method of receiving alert messages. The effectiveness of this monitoring is particularly important, not only due to the fact that pain is considered the fifth vital sign for representing basic bodily functions, health and quality of life, but also, due to its subjective nature.

Keywords: Chronic pain monitoring; pain diary; mobile health; web services; clinical decision support system;

1. Introduction

Pain is considered the fifth vital sign for representing basic bodily functions, health and quality of life,[1,2] complementing the well-known physiologic parameters of blood pressure, body temperature, pulse rate and respiratory rate. Nevertheless, it is distinguished from these vital signs, insofar it describes a subjective experience and manifests itself in a particular way in each individual. Actually, the pain relies of physiological, neurological and psychological idiosyncrasies. The International Association for the Study of Pain,[3,4] defines the pain as an unpleasant sensory and emotional experience related to past or potential tissue damage or it may be described through the concepts of tissue damage. When pain occurs quickly and with relatively short duration, is considered as acute pain. On the contrary, when pain manifests itself over a long period of time is regarded as chronic pain,[5] and may be related to a number of different pathological stages and medical conditions such as arthritis, fibromyalgia, migraine, low back pain, among others.

In fact, in accordance with Institute of Medicine (IOM),[6] only chronic pain, affects at least 116 million American adults (circa 37% of total population), surpassing the total affected by heart disease, cancer, and diabetes combined. In addition, the occurrence of chronic pain reduces the quality of life and impairs the working abilities of people,[7] culminating in a high cost, about 635 billion USD per year, in medical treatment and lost productivity. In this sense, chronic pain computerized monitoring systems become strategically important, in order to considerable improve benefits for patients and healthcare professionals (HCPs). For the health care system it can contribute to optimization of human and financial resources.

This chapter aims to explain a monitoring system,[8] with particular emphasis on the use of web services (WS), that enable the combination of pain diary with a personal health record (PHR).[9] This way, the following section presents electronic pain diaries, succeeding a section that describes WS concepts, and a section related to architecture of proposed system. Finally, are presented future trends and conclusions.

2. Pain Diaries

Since the chronic pain occurs over time, leads to a permanent need for the monitoring of patients by HCPs. In this sense, several daily measurements over a period of time are performed, in order to analyze the pain evolution and its relation to therapy defined by the HCP. These regularly collected data, yield pain diaries and making them a valuable means to assess a patients clinical course and to identify changes in health conditions.

Furthermore, it empowers the patients to actively contribute to their health care[10] as well as often providing pragmatic assistance such as medication record and medical appointment reminders.[11] Usually, the input data are based on self-reporting, observation, or even physiological data collected. However, due to the inherent subjectivity of pain it becomes difficult to determine the right treatments for the patient in which pain is manifested.[12] Thus, it is common practice to use rating scales and questionnaires as a means of measuring pain, such that the pain rating scales have a fundamental place in clinical practice.[13] The pain values can be entered individually or combined with other parameters, physiological or behavioral characteristics of patients, such as physical activity or eating habits.

Figure 1 depicts several types of pain scales, namely: Faces Pain Scale (FPS)[14,15] (the initial version contained 7 faces, but was subsequently adjusted to only 6), Numeric Rating Scale (NRS),[16] and Visual Analog Scale[17] (VAS).

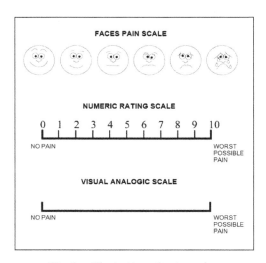

Fig. 1.   Illustration of pain scales.

In spite of FPS and VAS having been considered during the development of the computerized system, currently we adopted the NRS in the daily chronic pain software, in order to ask the patients to provide reports of their pain. The NRS ranges from 0 to 10, with the lower limit represents "no pain" and the upper limit represents the "worst pain imaginable". It can be stated simplistically that for values reported less than 2 is considered mild pain, for values between 3 and 7 is called moderate pain and for values between 8 and 10 and considered severe pain.

2.1. *Related Work*

Technology can provide several benefits including clinicians mobility, providing real-time access to data and information, reducing medical errors, saving time, supporting evidence-based practice, enhancing productivity and quality of care, and providing a tool for communication.[18] In this sense, the technological developments lead to pain diaries increasingly based on small, portable computers instead of using pencil-and-paper.[19] This way, the electronic pain diaries can be used to assist the patients in assessing and reporting their pain, and beyond that, can help the HCPs to deal with pain control in a more structured way.[20]

With this in mind, we developed an innovative system that uses WS in order to provide solutions to several limitations detected in literature, related to computerized pain diaries systems. Firstly, the electronic pain diary presented by Page *et al*,[21] consists in the software version of the McGill Pain Questionnaire (MPQ),[22] which runs in Microsoft XP Tablet-PC with exporting data capabilities to Microsoft Access. This approach exhibits two drawbacks, including the excessive time to complete the questionnaire (around 20 minutes, derived from the completion of 10 questions related not only with pain, but also with daily habits and symptoms), as well, the absence of real-time analysis by HCP in relation to recorded data.

For its part, Sufi *et al*[23] present a system to get the pain value based on mobile devices running software developed in Java 2 Micro Edition (J2ME). The value obtained is sent to a remote server, together with other physiological parameters such as heart rate or oximetry, using Short Message Service (SMS), Multimedia Messaging Service (MMS) or HyperText Transfer Protocol (HTTP). Nevertheless, an important limitation is observed, related to the nonexistent of schedule to patient's data acquisition, which may lead to forgetfulness by the patient, and therefore paucity or even absence of input pain records.

On the contrary, Ghinea *et al*,[24] present a client-server architecture whose clients are running in Windows CE handheld devices to gather patient's data around the clock. The collected information is sent to the server via an WiFi hotspot using HyperText Transfer Protocol Secure (HTTPS). However, this system presents a significant constraint, since it only sends data at the end of the day to the remote server, therefore, the analysis of data occurs with time lapse in order to the time of editing.

Finally, Bielli *et al*[25] present a pain diary based on mobile phones, whose pain information is sent to the server using a General Packet Radio Service (GPRS) connection or through web access. A peculiarity of the system is that it automatically sends SMS or MMS messages, to warn the patient to fill the required data. However, this approach presents a restriction regarding the obligation of data analysis by HCP before sending messages to the patients, *i.e.*, the system does not allow the generation of automatic responses, making it vulnerable to temporal availability of the HCP.

In summary, the presented system, as described below, takes advantage of WS features to provide real-time analysis and feedback, input data scheduling, and consequent adjustment of the therapy according to health conditions of each patient, along the treatment period. Moreover, this approach may lead to the adoption of WS as a means of integrating the patient's pain diaries in healthcare systems, thereupon, it may contributes to increase the interconnection among systems, and between HCPs and patients. Incidentally, the overwhelming percentage of smartphone downloadable pain management applications encountered in online marketplaces,[11] do not allow sending data to HCPs, neither integration with healthcare systems.

3. Web Services

The usage of WS have transformed the web from a publishing medium used to simply disseminate information, into an ubiquitous infrastructure that supports transaction processing.[26] The main purpose is to ensure interoperability, in other words, the WS provide a standardized mechanism for heterogeneous information systems and applications to communicate with each other. Furthermore, they are used to enable the reuse of application-components, and also to connect existing software, independent of their implementation language, operating platform,[27,28] and location.

The WS involves the presence of a provider, in charge for the service implementation and it availability on the Internet and a client to consume the service. Figure 2 depicts the WS protocol stack, composed by the following elements: discovery, description, messaging, and transport.

UDDI	Discovery
WSDL	Description
XML, SOAP	Messaging
HTTP, SMTP, FTP	Transport

Fig. 2.   Protocol stacks of web services.

The discovery layer comprises by the Universal Description Discovery and Integration (UDDI),[29] in order to provide a technical specification for describing and discovering WS providers, as well as their available services. In its turn, the description layer is composed by Web Service Description Language (WSDL),[30] that consists in the definition of the public interface to the WS, in terms of Extensible Markup Language (XML) syntax. The obtained information contains the name, location, the operations exhibited by the WS, and expected inputs and outputs. The messaging layer is responsible for encoding and exchanging data between provider and client. In this sense, is often used the XML, and the Simple Object Access Protocol (SOAP).[26,27] Finally, the transport layer supports several protocols such as HTTP, Simple Mail Transfer Protocol (SMTP), and File Transfer Protocol (FTP), among others.

3.1. *XML*

XML defines documents in a structured format such as data content and metadata, that enables to exchange information among different computer systems independently of their platform and environment. This structure is composed by labels specified in a tag format, that represent the scheme and the content regarding to the data. Each label is described by a pair of tags, such as <> and </>, that identify respectively the start and the end of the data. The start tag may include a name-value pair termed *attribute* in order to typify the content of the label. These labels represent a portion of the document and are denominated *element*. In its turn, the elements are grouped into a hierarchical structure by defining parent-child relationships. The top-level element is called *document root* and is unique in the XML tree. An example of the XML structure is depicted in Figure 3.

```
<? xml version="1.0" enconding="UTF-8" ?>
<rootelement>
    <element attribute_name="attribute_value">
        <child>This is level 1 of the element</child>
    </element>
</rootelement>
```

Fig. 3.   Example of the XML structure.

3.2. *SOAP*

SOAP is a lightweight protocol that grants an extensible XML framework for message exchange, over a different transport protocols, usually HTTP, in a distributed environment. In fact, it is based on XML and specifies a manner to exchange messages between different processes and/or machines. This specification is called *envelope*, and it is the root element of the SOAP Message, which purpose is to define the origin, the destination, and the process model through the use of XML to encode data types contained in the messages. The *message path* is the set of intermediates processes through which the message passes since the origin to the destination.

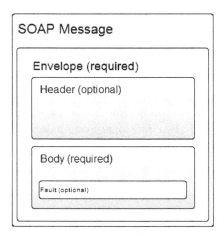

Fig. 4.   SOAP message structure.

The SOAP Envelope, depicted in Figure 4, provides the serialization context and namespace information for data handled in the message, and is comprised by the following elements:

- SOAP Header: Contains the required information about the body content processing, such parameters regarding routing, delivery, authentication and authorization. This is an optional element in SOAP Envelope.
- SOAP Body: Is a mandatory element that includes the data, expressed in terms of XML, to be processed and delivered. Optionally, the body can include the fault element, in order to display error messages.

4. System Architecture

The presented approach encompasses a commercial PHR, called Meu Sapo Saúde, provided by PT Comunicações/SAPO Labs, and a mobile application (app) used by the patients as pain diary. Both PHR's module of pain and the app were developed within this research, and are connected through the use of WS. The adoption of WS was due to the fact that they provide the usability and interoperability required to ensure the integration of pain diary records in a remote database associated to PHR. The app was developed for devices with Android OS and includes a SQLite database to store local data. The workflow of the system, depicted in Figure 5, is described as follows.

Firstly, (1) HCPs, using a browser, access to the PHR to define the monitoring plan of each patient in terms of frequency of recorded values and content of automatic messages based on obtained values. (2) This way the app, due to the fact that periodically checks for updates in the PHR, changes the monitoring rules in order to adjust them in agreement with the clinician's indications. (3) The app saves these data internally in a SQLite database. Therefore, over time the individual therapy of each patient tends to remain adjusted according to the evolution of his state of health.

With this in mind, (4) in conformity with the frequency of data recording in the diary of pain defined by the HCP, the system asks the patient to enter the pain data. This request is followed by an audible warning and remains on the mobile device's screen over a period of time. After this period, if the patient has not responded, a "no response" is assumed, which will then be statistically analyzed together with other values. Whether a "no response" or a value are entered by the patient, they are (5) immediately recorded in the database of the mobile device, as well as (6) being sent by a WS to the PHR, thereby available for online viewing. If the data transmission is not successful, the records will be marked as pending and the system will try again to send them the next time planned for recording data. (7) Automatically and without requiring intervention by the patient, the system ensures the sending of all data to the PHR and therefore allows a reliable data analysis. (8) Immediately after sending and recording the values of pain, the app will go into background mode until the next moment of data entry. In addition, (9) the app periodically detects, through the WS, the existence of messages in the PHR. These messages may have been caused by (10) the last data recorded or (11) issued by an HCP. (12) Whenever there are messages, they are saved in local database of the mobile device and are presented to the patient. If the app is in background, its activation is following by an audible warning.

Furthermore, (13) the system allows the patient to register unplanned pain records in which submission process is identical to the planned records. These data are classified according to their nature, *ie*, for analytical purposes each record indicates if it was planned or unplanned. (14) This register of unplanned data can be performed directly in the PHR, by using a browser. At last, (15) all the information generated in the system, such as pain records and alert messages, can be accessed in the PHR, through the use of the browser, either by the patient or by an HCP.

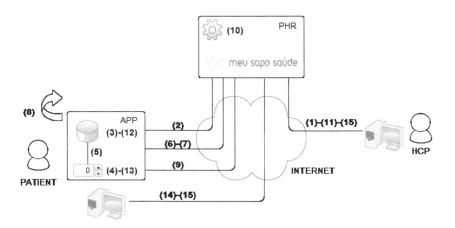

Fig. 5. Workflow of the proposed system.

Taking into account the abovementioned processing, the use of the WS, through Internet access allows the user to take advantage of the mobile device's ubiquity and connectivity. In fact, WS enables communication between the app and the PHR, which consists of the execution of several methods, namely:

- Scheduling: Get the data entry frequency around-the-clock. This frequency vary according to the health of the patient;
- Messaging: Obtain messages for the patient. These messages were issued manually by the HCP or automatically by the system;
- Pain Records: Sends the pain records emitted by the patient. The pain records vary between planned and unplanned.

The app sends SOAP messages over HTTP using a standard transport security, such as HTTPS to ensure that a message is protected during transit. In other words, the HTTPS is a *point-to-point* security, which does not allow intermediaries to act on the data, and requires trust between the HTTPS end-point and the location of the application being secured.[31] In order to inform the app that a message has reached its destination the WS sends a response whose format can either be SOAP or JavaScript Object Notation (JSON).[32] This request-response implementation is called two-way callback-based asynchronous send.[33] The Figure 6 depicts a request and a response in SOAP format, regarding to obtain the pending messages of the patient.

It should be noted that the WS associated with the PHR was developed on Microsoft technology, particularly by using Windows Communication Foundation (WCF).[28,34]

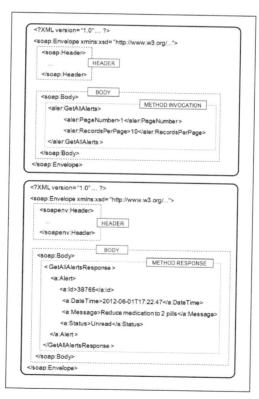

Fig. 6.   Example of SOAP request-response message.

In summary, the system presents an easy access to the patient, since it happens not only through the app, but also directly in the PHR. At the same time, is provided a two-way communication between the patient and HCP, to the extent that the data recorded by the first can trigger the issuance of warnings pre-defined by the second. Furthermore, the automation of messages emission will release the HCP's time spent in data analysis and therefore solve one common problem related to the lack of regularity in the visualization and incorporation of obtained data in decision making by the HCPs.[35] Moreover, due to the use of WS, the feedback under normal conditions occurs in real-time. This feature may lead to faster and immediately adjust of the medical procedures after the occurrence of an episode of pain. Additionally, the system allows the patient to register unplanned pain records whenever there is an occurrence of pain. Thus, the monitoring data will be more comprehensive and realistic about the patient's state of health and consequently may result in a higher effectiveness of the therapy defined by the HCP. This way, the user's experience resulting from the interaction with the system will be enhanced, which may lead to increase the adherence of patients.

5. Conclusion

In this paper it was presented the use of WS in order to enhance the features of pain diaries, especially with respect to monitoring and implementation of clinical practice by the HCP. The results obtained in the pilot study are very promising and reveal that this approach, mainly due to the use of WS, allows to solve several problems detected in different papers and reviews. These problems include the lack of timely feedback from the HCP or the adjustment of the system depending on the patient's treatment. Due to the detection and retrieval of messages through the use of WS, it is guaranteed that the patient is alerted in a timely manner with warning messages defined in the system or manually issued by the HCP. Besides, since the system allows the definition of automatic responses according to the values obtained for the pain, it does not require the permanent expenditure of time by HCPs in analyzing and formulating responses. Moreover, the system determines the behavior of the pain diary in terms of frequency of records and display of alert messages, making it an adjustable system to the patient and their therapy.

However, new studies should be addressed to confirm these evidences, so that the system will be deployed in several Hospital Centers to cover a wide range of patients. During this implementation numerous studies should be performed by a multidisciplinary team of experts, in order to evaluate this system. It should be appraised the usability (of the app and the PHR), economic effects, and the contribution to improve the patient's treatments adherence and the effectiveness of the therapeutics. In this sense, the present system will be complemented with a knowledge based component whose purpose is to analyze and to process the obtained patients' pain records.

References

1. M. McCaffery and C. L. Pasero, *The American journal of nursing* **97**, 15 (1997).
2. M. K. Merboth and S. Barnason, *The Nursing clinics of North America* **35**, 375 (2000).
3. H. Merskey and N. Bogduk, *Classification of Chronic Pain: Descriptions of Chronic Pain Syndromes and Definitions of Pain Terms* (International Association for the Study of Pain, 1994), pp. 209–214.
4. J. D. Loeser and R.-D. Treede, *Pain* **137**, 473 (2008).
5. A. V. Apkarian, M. N. Baliki and P. Y. Geha, *Progress in Neurobiology* **87**, 81 (2009).
6. C. Committee on Advancing Pain Research and E. I. of Medicine, *Relieving Pain in America: A Blueprint for Transforming Prevention, Care, Education, and Research* (The National Academies Press, 2011).
7. M. A. Ashburn and P. S. Staats, *The Lancet* **353**, 1865 (1999).
8. N. Pombo, P. Araújo, J. Viana, B. Junior and R. Serrano, Contribution of web services to improve pain diaries experience, in *Lecture Notes in Engineering and Computer Science: Proceedings of The International MultiConference of Engineers and Computer Scientists*, , IMECS 2012 Vol. 2195(1)2012.
9. M. Wang, C. Lau, I. Matsen, F.A. and Y. Kim, *Information Technology in Biomedicine, IEEE Transactions on* **8**, 287(sept. 2004).
10. J. Gaertner, F. Elsner, K. Pollmann-Dahmen, L. Radbruch and R. Sabatowski, *Journal of Pain and Symptom Management* **28**, 259 (2004).
11. B. A. Rosser and C. Eccleston, *Journal of Telemedicine and Telecare* **17**, 308 (2011).
12. J. Giordano, K. Abramson and M. Boswell, *Pain Physician* **13**, 305 (2010).
13. A. Williamson and B. Hoggart, *Journal of Clinical Nursing* **14**, 798 (2005).
14. D. Bieri, R. A. Reeve, G. Champion, L. Addicoat and J. B. Ziegler, *Pain* **41**, 139 (1990).
15. C. L. Hicks, C. L. von Baeyer, P. A. Spafford, I. van Korlaar and B. Goodenough, *Pain* **93**, 173 (2001).
16. E. Joos, A. Peretz, S. Beguin and J. P. Famaey, *The Journal of rheumatology* **18**, 1269 (1991).
17. M. D. Miller and D. G. Ferris, *The Family practice research journal* **13**, 15 (1993).
18. Y.-C. Lu, Y. Xiao, A. Sears and J. A. Jacko, *International Journal of Medical Informatics* **74**, 409 (2005).

19. M. Morren, S. van Dulmen, J. Ouwerkerk and J. Bensing, *European Journal of Pain* **13**, 354 (2009).

20. L. Lind, D. Karlsson and B. Fridlund, *International Journal of Medical Informatics* **77**, 129 (2008).

21. D. B. Page, F. Weaver, D. J. Wilkie and T. Simuni, *Parkinsonism & Related Disorders* **16**, 139 (2010).

22. R. Melzack, *PAIN* **1**, 277 (1975).

23. F. Sufi, Q. Fang and I. Cosic, A mobile phone based intelligent scoring approach for assessment of critical illness, in *Information Technology and Applications in Biomedicine, 2008. ITAB 2008. International Conference on*, may 2008.

24. G. Ghinea, F. Spyridonis, T. Serif and A. Frank, *Information Technology in Biomedicine, IEEE Transactions on* **12**, 27(jan. 2008).

25. E. Bielli, F. Carminati, S. La Capra, M. Lina, C. Brunelli and M. Tamburini, *BMC Medical Informatics and Decision Making* **4**, p. 7 (2004).

26. J. Tekli, E. Damiani, R. Chbeir and G. Gianini, *Services Computing, IEEE Transactions on* **PP**, p. 1 (2011).

27. K. Y. Lai, T. K. A. Phan and Z. Tari, Efficient soap binding for mobile web services, in *Local Computer Networks, 2005. 30th Anniversary. The IEEE Conference on*, nov. 2005.

28. W. Zhang and G. Cheng, A service-oriented distributed framework-wcf, in *Web Information Systems and Mining, 2009. WISM 2009. International Conference on*, nov. 2009.

29. T. Bellwood, S. Capell, L. Clement, J. Colgrave, M. J. Dovey, D. Feygin, A. Hately, R. Kochman, P. Macias, M. Novotny, M. Paolucci, C. von Riegen, T. Rogers, K. Sycara, P. Wenzel and Z. Wu, UDDI Version 3.0.2 http://uddi.org/pubs/uddi_v3.htm (October 2004).

30. A. A. Lewis, *Interface*, 1 (2007).

31. C. A. Ardagna, E. Damiani, S. D. C. di Vimercati and P. Samarati, *Electronic Notes in Theoretical Computer Science* **142**, 47 (2006).

32. G. Wang, Improving data transmission in web applications via the translation between xml and json, in *Communications and Mobile Computing (CMC), 2011 Third International Conference on*, april 2011.

33. J. Kangasharju, T. Lindholm and S. Tarkoma, *Computer Networks* **51**, 4634 (2007).

34. M. Youxin, W. Feng and Z. Ruiquan, *Software Engineering, World Congress on* **4**, 100 (2009).

35. L. D. Marceau, C. L. Link, L. D. Smith, S. J. Carolan and R. N. Jamison, *Journal of Pain and Symptom Management* **40**, 391 (2010).

PARALLEL BINOMIAL AMERICAN OPTION PRICING ON CPU-GPU HYBRID PLATFORM

NAN ZHANG

Department of Computer Science and Software Engineering,
Xi'an Jiaotong-Liverpool University, Suzhou, China
E-mail: nan.zhang@xjtlu.edu.cn

CHI-UN LEI

Department of Electrical and Electronic Engineering,
The University of Hong Kong, Hong Kong
E-mail: culei@eee.hku.hk

KA LOK MAN

Department of Computer Science and Software Engineering,
Xi'an Jiaotong-Liverpool University, Suzhou, China
Myongji University, South Korea
Baltic Institute of Advanced Technology, Lithuania
E-mail: ka.man@xjtlu.edu.cn

We present a novel parallel binomial algorithm to compute prices of American options. The algorithm partitions a binomial tree into blocks of multiple levels of nodes, and assigns each such block to multiple processors. Each processor in parallel with the others computes the option's values at the assigned nodes. The algorithm is implemented and tested on a heterogeneous system consisting of an Intel multi-core processor and a NVIDIA GPU. The whole task is split and divided over the CPU and GPU so that the computations are performed on the two processors simultaneously. In the hybrid processing, the GPU is always assigned the last part of a block, and makes use of a couple of buffers in the on-chip shared memory to reduce the number of accesses to the off-chip device memory. The performance of the hybrid processing is compared with an optimised CPU serial code, a CPU parallel implementation and a GPU standalone program.

Keywords: Parallel computing; Option pricing; Binomial method; Graphics processing unit; Heterogeneous processing

1. Introduction

An American call/put option is a financial contract that gives the contract buyer the right, but not the obligation, to buy/sell at a strike price K a unit of certain stock, whose current price is S_0, at any time until a future expiration date T. If the buyer of the contract chooses to exercise the right the option seller must sell/buy a unit of the stock to/from the buyer at the strike price. Since such a contract gives the buyer a right without any obligation, the buyer of the contract must pay the seller a certain amount of premium for this right. The problem of option pricing is to compute the fair price of the contract to both the seller and the buyer.

Black, Scholes and Merton studied this problem, and published their work[1,2] in 1973. They deduced closed-form formulae for calculating the prices of European call and put options. These options can only be exercised at the expiration date T. However, for American options, because of the early exercise feature, no closed-form pricing formula has been found. Instead, their price must be computed using numerical procedures, such as the binomial methods.

Option pricing is a crucial problem for many financial practices and so is to be completed with minimal delay. Nowadays, as parallel computers become widely available, many new developments have been advanced in applying parallel computing to the problem of option pricing. Some researchers developed parallel algorithms for various option pricing problems on shared- and distributed-memory multi-processor computers,[3,4] and some developed algorithms for option pricing on GPUs.[5–7]

In this paper, we present a parallel algorithm for pricing American options on a heterogeneous system hosting both a shared-memory multi-core processor and a NVIDIA GPU. The algorithm computes prices of American options on recombining binomial trees. The computation is split and divided over the CPU cores and GPU. The implementation was tested on a laptop system with an Intel dual-core P8600 and a NVIDIA Quadro NVS 160M. The performance of the algorithm was tested and analysed.

2. Related work

Gerbessiotis[3] presented a parallel algorithm that computes the price of a European (or an American) option on a recombining binomial tree. The algorithm partitions a binomial tree into blocks of multiple levels. Each block is further divided and assigned to distinct processors. A processor partitions the sub-block of nodes that has been assigned to it into two regions. Computation in one of the two regions depends on results from nodes out of the region, while that in the other does not have such external dependency. In such a scheme, each block is processed in parallel by multiple processors. The assignment of sub-blocks to processors is fixed from the beginning of the computation. The performance of the algorithm is analysed following the bulk-synchronous parallel model. The implementation of the algorithm is tested in a cluster of PC workstations under a message-passing interface (MPI) and a non message-passing interface.

While the above algorithm parallelises a binomial tree along the axis that represents the stock's price, Ganesan et al.[8] presented an algorithm where the processing of a binomial tree was parallelised along the time-axis. The algorithm was implemented on a GeForce 8600GT GPU. Since the way they implemented the algorithm and the 16KB shared memory limitation, the implementation can only work with a tree of maximumly 1024 time steps.

Solomon et al.[5] presented a GPU-based trinomial algorithm for pricing European options and a binomial algorithm for pricing American look-back options. The algorithms were implemented and tested on a NVIDIA GTX260 GPU using the CUDA[9] programming model. In pricing the American look-back option on a binomial lattice, the authors implemented a hybrid method where a threshold was pre-set. As the backward computation proceeds from the leaf nodes to the root, the level of parallelism reduces during the course. So in their algorithm when the backward induction passes over the threshold the computation was taken over by the CPU. The assumption was that the CPU could likely perform the later calculations faster than the GPU.

For CPU and GPU to work side by side for hybrid parallel processing several challenges have to be solved. Tomov et al.[10] discussed ideas in this respect with the development of a hybrid LU factorisation algorithm where the computation was split over a multi-core and a graphics processor. Of such challenges the synchronisation between CPU and GPU or even between different thread blocks of a GPU is foremost. At the moment NVIDIA's CUDA programming environment does not provide any means by which inter-block coordination can be easily handled. So when algorithms are designed for NVIDIA GPUs, the computation task must be decomposed in such a way that each thread block is executed independently. This restriction has caused problems to the applications where explicit inter-block synchronisation has to be employed. Some researchers have been looking into this issue. For example, Xiao et al.[11] developed three inter-block synchronisation schemes for NVIDIA GPUs. As in the CUDA programming model the execution of a thread block is non-preemptive, in the schemes, they use an one-to-one mapping between thread blocks and multi-processors (also known as streaming processors). However, a very simple solution to this problem is to stop the GPU kernel at a point where synchronisation is needed and later re-start it. But this stopping and re-launching of GPU kernels is a high-costly and hurts performance. This is the synchronisation method we adopted in the implementation of our hybrid algorithm.

3. Binomial American option pricing

The binomial tree model[12] is a widely-used numerical solution to various problems in computational finance. A recombining binomial tree models the dynamic change of a stock price within the time frame from 0 to T. For a binomial tree of N time steps there are $N + 1$ node levels, corresponding to the $N + 1$ time spots where $t = 0, 1, 2, \ldots, N$. Any interior node (for example, denoting stock price S) has two successors – an up-move node (denoting price uS) and a down-move node (denoting price dS). If the annual volatility of the stock price is σ, we set u to be $u = \exp(\sigma\sqrt{T/N})$ and $d = 1/u$, so that the tree describes the discrete version of the continuous price change. An example of a recombining binomial tree of 3 time steps and 4 levels is shown in Fig. 1.

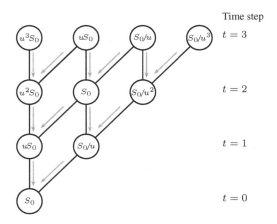

Fig. 1. A recombining binomial tree of 3 time steps and 4 levels. The price at each node is shown in the node. Note that in such a tree at any level $t = n$, the number of nodes in that level is $n + 1$.

To compute the price of an American option (expiration T, strike price K) on a stock (current price S_0), assuming annual continuous compound interest rate R, we start by calculating the option's payoff P_T ($\max(S_T - K, 0)$ for a call, and $\max(K - S_T, 0)$ for a put) at each leaf node. We set the option's value π_T at each leaf node to be the option's payoff P_T at that node. For an interior node (assuming the price at which is S_t) we calculate the discounted expected option value $r^{-1}\mathbb{E}(\pi_{t+1}|S_t)$ at the node S_t as $r^{-1}\mathbb{E}(\pi_{t+1}|S_t) = r^{-1}(p\pi_{t+1}^u + (1-p)\pi_{t+1}^d)$, where $r = \exp(RT/N)$ is the one time-step interest rate, $p = (r - d)/(u - d)$ is the risk-neutral probability of the up-move, and π_{t+1}^u and π_{t+1}^d are the option's values at the successive up-move and down-move nodes, respectively. Then we set the option's value π_t at node S_t to be the maximum of the discounted expectation and the immediate payoff P_t. So we have $\pi_t = \max(P_t,\ r^{-1}\mathbb{E}(\pi_{t+1}|S_t))$. We apply these steps to all the interior nodes of the binomial tree in the backward induction manner until we get π_0 at the root node.

4. The CPU-GPU heterogeneous system

The hardware platform (Fig. 2) we used in our work was a laptop system that equipped with a dual-core Intel P8600 (2.4GHz) and a NVIDIA Quadro NVS 160M. The NVIDIA GPU has a single multi-processor that integrates 8 CUDA cores. Their clock speed is 1.45GHz. On-chip the graphics processor has 8KB registers and 16KB shared memory. Off-chip the processor has 256MB device memory installed, which serves as the local, global, constant memories, etc. Accessing the on-chip shared memory is much faster than accessing the device memory. According to NVIDIA's manual[9] the Quadro NVS 160M is of compute capability 1.1, and so it only supports single-precision floating point arithmetic. Eight single-precision floating point operations can be performed per clock cycle per multi-processor.

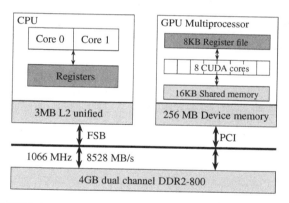

Fig. 2.   The CPU-GPU heterogeneous system with an Intel P8600 and a NVIDIA Quadro NVS 160M.

5. The parallel American option pricing algorithm

To compute the price of an American option on a binomial tree of N time steps ($N + 1$ time spots) the parallel algorithm partitions the tree into blocks of multiple levels of nodes. Each block is further divided into equal-sized (except the last one) sub-blocks. The blocks are processed in a sequential order backwards from the leaf nodes. However, within each block the sub-blocks are processed in parallel by distinct processors. The parallel processing of a block consists of two phases. In phase one each processor computes at the half (region A) of the sub-block which has no dependency on nodes out of the region. Once all the processors finish computing nodes in their region A, phase two begins in which each processor computes at the nodes in the remaining half (region B) of the sub-block. After phase two is completed, all the processors move onto the next block. The parallel processing on a binomial tree of 11 time steps is shown in Fig. 3.

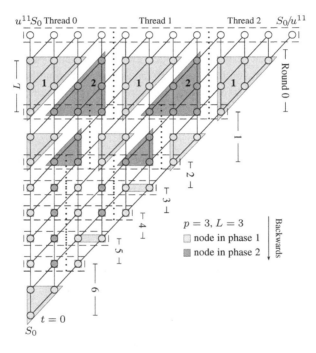

Fig. 3.   The parallel algorithm on a binomial tree of 11 time steps.

In the algorithm we defined a parameter L which specifies the maximum number of levels that a block can have. However, the actual number of levels in a block is also determined by the number of nodes that each processor gets in the current base level. To save all the intermediate results each processor maintains a local buffer. The buffer is of $(L + 1)$ rows and $(N + 2)$ columns. To avoid excessive memory transactions the buffer is used in a modulo wrapping around manner. Fig. 4 shows such an example where in round 0 the base level nodes are saved in row 0 of the buffer, and after the block is processed the nodes saved in row 3 become the base level nodes of the next round. At this point, they are not copied back to row 0.

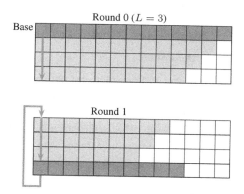

Fig. 4. The modulo wrapping around manner of the local buffer kept by each processor.

The synchronisation scheme used in the algorithm is shown in Fig. 5. It is always the case that the last thread has no node to process in phase two.

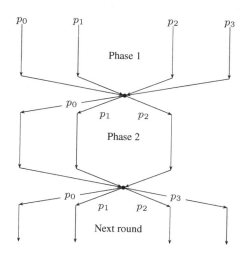

Fig. 5. The synchronisation between multiple threads.

For a p-way parallelism (p distinct processors are used) on an N-step binomial tree, because processor p_0 roughly processes $1/p$ of the total nodes in the tree, the parallel runtime $T_P = O(N^2/p)$, the parallel speedup $S = T_S/T_P = O(p)$ (T_S is the serial runtime), the parallel efficiency $E = S/p = O(1)$, and the cost pT_P of the parallel algorithm is $pT_P = O(N^2)$. So the parallel algorithm is cost-optimal in that the cost has the same asymptotic growth rate as the serial case.

6. Performance testing on the P8600

On the dual-core Intel P8600 we implemented the parallel algorithm and compared its performance against an optimised serial implementation of the pricing algorithm. We used an American put option in the tests, where the parameters were set as: current stock price $S_0 = 100$, strike price $K = 100$, option expiry date $T = 0.6$, annual continuous compound interest rate $R = 0.06$ and annual volatility $\sigma = 0.3$. The number N of time steps varied from 4×10^3 to 56×10^3, with an increment of 4000 in each test. In the parallel implementation we used two threads and explicitly bound them onto each of the two cores of the CPU, as Fig. 6 shows.

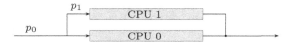

Fig. 6.   Binding two threads onto the two cores of the CPU.

In computing the stock's price at a node we did not use S_0. We know that the stock's price S_n^0 at the first node (at column 0) of a certain level where $t = n$ is, as it is shown in Fig. 1, $S_n^0 = u^n S_0$. So the price S_n^j at the j-th column in that level is $S_n^j = u^{-2j} S_n^0$. So for the nodes in any level $t = n$ we computed S_n^0 once and re-used its value at the remaining nodes of the level. By this way we avoided repeatedly evaluating the same mathematical expression. This optimisation made noticeable improvement to the performance of the implementation.

We used single-precision floats in the programs so that we could make comparisons between the performances on the CPU and on the GPU. The operating systems used was Ubuntu Linux 10.10 (64-bit version). The compiler used was Intel's icpc 12.0 for Linux with optimisation options -O3 and -ipo switched on. The POSIX thread library used was NPTL (native POSIX thread library) 2.12.1. The parallel speedups in all the tests are plotted in Fig. 7. In the tests we observed super-linear speedup in some of the tests. This must have been caused by the caching effect and the more efficient use of the system bus. In this group of tests the maximum number L of levels in a block was set to 20.

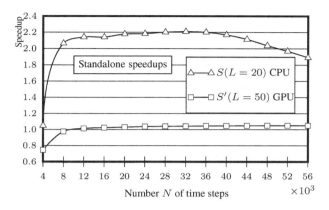

Fig. 7.   The speedups of the CPU parallel implementation and the GPU implementation.

7. The GPU algorithm and its performance

Programming the same binomial American option pricing problem on the Quadro NVS 160M is very different from working with the Intel P8600, because of the SIMT (single instruction multiple threads) execution model of the NVIDIA GPU. The NVIDIA CUDA 4.0 SDK comes with an example where thousands of European calls are priced using the binomial method.[13] In the example, a single one-dimensional thread block is used to price a single call option. To avoid frequent access to the off-chip global memory but to make use of the on-chip shared memory as much as possible, the algorithm partitions a binomial tree into blocks of multiple levels. The partition pattern is very similar to the one shown in Fig. 3, except that the NVIDIA's algorithm requires that all the blocks have the same number of levels and this number must be a multiple of two. The algorithm also uses two buffers in the shared memory. The algorithm begins by allocating an one-dimensional buffer in the global memory. All the threads in the thread block compute the option's payoffs at the leaf nodes and save them into the buffer. When processing a block of interior nodes, the threads first load the computed option values from the global buffer into one of the two shared buffers. Then the computation is carried out between the two shared buffers. After this the results are copied back to the appropriate positions in the global buffer. The threads then move to the next part of the block to repeat the same processing.

The algorithm we implemented on the Quadro NVS 160M modified the NVIDIA's algorithm by allowing arbitrary number of levels in a block. A run of our algorithm for NVIDIA GPUs is shown in Fig. 8.

The performance comparison between this GPU implementation and the CPU

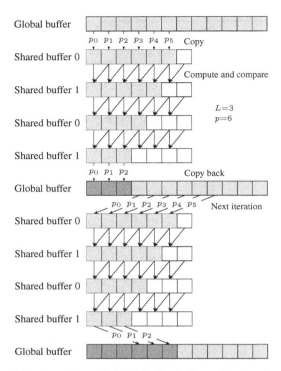

Fig. 8.   GPU binomial option pricing with double buffers in the on-chip shared memory. Note that in this example we have 6 threads.

sequential program is plotted in Fig. 7, where the same American put option example was used. From the results we can see that the performance on the GPU was almost the same (or, slightly better in some cases) as that on a single core of the CPU. Without the double-buffer memory access optimisation the GPU's performance was far worse. In all the GPU tests the parameter L (the maximum number of levels in a block) was set to 50, much increased from the CPU parallel tests where L was 20. This was to reduce the number of times that the GPU threads have to access the buffer in the global memory.

8. The CPU-GPU hybrid processing

To compute the parallel binomial algorithm (Fig. 3) using both the CPU and the GPU in the laptop system (Fig. 2) we assigned the GPU the last sub-block (for example, the part processed by thread p_2 in Fig. 3) in each round, because the GPU algorithm is not suitable for a sub-block that has nodes in region B. Moreover, as the GPU's performance (Fig. 7) on this pricing problem was almost identical to a single core of the CPU, initially the workload was divided equally among the two cores of the CPU and the GPU.

To coordinate the GPU with the two cores of the CPU we have to use one of the two cores for the communication and the synchronisation. Since the launch of a kernel on the GPU is asynchronous, that is, control is returned to the CPU before the task on the GPU is completed, we did not leave the coordinating core of the CPU idle while the kernel is executed on the GPU. We assigned an equal part of the total workload to the coordinating core. This distribution of workload on the CPU and GPU is illustrated in Fig. 9.

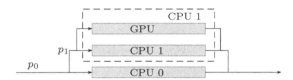

Fig. 9. A CPU core is used to coordinate with the GPU, while an equally-sized workload is assigned to that core.

The parallel algorithm (Fig. 3) requires all the threads working at block i to finish before the processing of block $i + 1$ to start. So at the end of each round the GPU kernel had to be ended and a new kernel was launched at the start of each new round. Launching a new kernel every round is a high-costly operation, but this is the price that one has to pay in order to make the CPU and the GPU work side by side. Algorithm 1 shows the steps performed by this coordinating core of the CPU.

To see the performance of the hybrid algorithm we did two groups of tests where L, the maximum number of levels in a block, was set to 20 and 50, respectively. The tests were made using the same American put option with the same parameter setting. The GPU kernel were launched with a single thread block of 512 threads. The speedups are plotted in Fig. 10.

From the results we can see that when $L = 20$ the CPU parallel implementation with 2 working threads outperformed the hybrid processing, but when $L = 50$ the opposite situation was observed. The reason for the first observation was that the repetitive launching of GPU kernels reduced the performance of the hybrid processing. When $L = 50$, the number of launchings was reduced so that the performance of the hybrid processing was improved. However, when $L = 50$ the CPU parallel code became poorly performed. The reason was that the local buffer that saved the intermediate results at the CPU side became less efficient for

---

Algorithm 1: Computational steps performed by the coordinating core.

---

begin
    // Initialisation
1   Compute option's payoffs at the end-level nodes assigned to the core and the GPU;
    // Backward induction
2   **while** *there is a next round* **do**
        // Phase 1
3   **if** *GPU is needed* **then**
4   Launch kernel for the part assigned to the GPU;
5   Compute at region A of the sub-block assigned to the CPU core;
6   **if** *GPU is needed* **then**
7   Wait for the GPU to finish;
8   Copy data from the GPU buffer to the CPU buffer;
9   Synchronise with the other CPU cores;
        // Phase 2
10   **if** *there is region B* **then**
11   Compute at region B of the sub-block assigned to the CPU core;
12   **if** *GPU is needed* **then**
13   Copy data from the CPU buffer to the GPU buffer;
14   Synchronise with the other CPU cores;
        // For next round
15   Update variables and parameters;
end

---

caching when L became large. According to the theoretical analysis the speedup S of this parallel algorithm is $S = O(p)$. However, the performance of the hybrid processing after adding the GPU did not show significant enhancement over the CPU parallel code. We believe that this was due to the coordination overhead between the CPU and the GPU.

9. Conclusion

We have presented a parallel algorithm that computes the price of an American option on a recombining binomial tree. The tree is partitioned into blocks of multiple levels of nodes. A block is divided into sub-blocks and these sub-blocks are assigned to distinct processors to be processed in parallel. The processing of a block by multiple processors consists of two phases. In phase one, the processing is carried out on the nodes at which the computation has no external dependency, and in phase two, the nodes are processed where such dependency exists and has been resolved in phase one. The parallel algorithm dynamically adjusts the assignment of sub-blocks to processors since the level of parallelism decreases as the computation proceeds from the leaf nodes to the root.

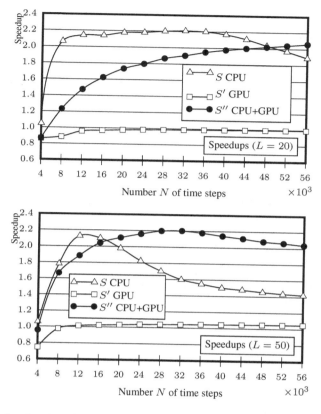

Fig. 10. Speedup plots of the CPU parallel implementation and the hybrid implementation.

The parallel algorithm was implemented on the dual-core CPU. In some of the test cases super-linear speedups were observed against an optimised serial CPU code. A GPU binomial pricing algorithm was then discussed, where double buffers in the on-chip shared memory are used to reduce the number of accesses to the off-chip device memory.

The parallel algorithm was then adapted to the dual-core CPU and the GPU. The partition of a binomial tree is in such a way that the GPU is always given the last sub-block to compute. To coordinate the GPU with the CPU we had to use one of two CPU cores to repeatedly launch the GPU kernel and then stop it at the synchronisation point. This has caused much overhead and reduced the performance of the hybrid processing.

174

Acknowledgment

This research work is jointly sponsored by Transcend Epoch International Co., Ltd - Belize and Hong Kong, by the XJTLU Research Development Fund Grant No. 10-03-08 and by HKU Seed Funding Programme for Basic Research Grant No. 201111159068. Previous conference paper appeared in IAENG IMECS 2012.[14]

References

1. F. Black and M. Scholes, The Pricing of Options and Corporate Liabilities, *The Journal of Political Economy* **81**, 637 (1973).
2. R. Merton, Theory of Rational Option Pricing, *Bell Journal of Economics and Management Science* **4**, 141 (1973).
3. A. V. Gerbessiotis, Architecture Independent Parallel Binomial Tree Option Price Valuations, *Parallel Computing* **30**, 301 (2004).
4. M. Zubair and R. Mukkamala, High Performance Implementation of Binomial Option Pricing, *Lecture Notes in Computer Science* **5072**, 852 (2008).
5. S. Solomon, R. K. Thulasiram and P. Thulasiraman, Option Pricing on the GPU, in *Proceedings of the 12th IEEE International Conference on High Performance Computing and Communications*, (Melbourne, Australia, 2010).
6. B. Dai, Y. Peng and B. Gong, Parallel Option Pricing with BSDE Method on GPU, in *Proceedings of the 9th International Conference on Grid and Cloud Computing*, (Nanjing, China, 2010).
7. V. Surkov, Parallel Option Pricing with Fourier Space Time-stepping Method on Graphics Processing Units, *Parallel Computing* **36**, 372(Jul 2010).
8. N. Ganesan, R. D. Chamberlain and J. Buhler, Accelerating Options Pricing Calculations via Parallelization along Time-axis on a GPU, in *Proceedings of the 1st Symposium on Application Acceleration and High Performance Computing (SAAHPC '09)*, (Urbana-Champaign, Illinois, 2009).
9. NVIDIA Corporation, *NVIDIA CUDA C Programming Guide (version 4.0)*, (2011).
10. S. Tomov, J. Dongarra and M. Baboulin, Towards Dense Linear Algebra for Hybrid GPU Accelerated Manycore Systems, *Parallel Computing* **36**, 232(Jun 2010).
11. S. Xiao and W. chun Feng, Inter-block GPU Communication via Fast Barrier Synchronization, in *Proceedings of 2010 IEEE International Symposium on Parallel & Distributed Processing (IPDPS)*, (Atlanta, GA, 2010).
12. J. C. Cox, S. A. Ross and M. Rubinstein, Option Pricing: A Simplified Approach, *Journal of Financial Economics* **7**, 229(Sep 1979).
13. C. Kolb and M. Pharr, Options Pricing on the GPU, in *GPU Gems 2: Programming Techniques for High-Performance Graphics and General-Purpose Computation*, eds. M. Pharr and R. Fernando (Addison-Wesley, 2005)
14. N. Zhang, E. G. Lim, K. L. Man and C.-U. Lei, CPU-GPU Hybrid Parallel Binomial American Option Pricing, in *Lecture Notes in Engineering and Computer Science: Proceedings of The International MultiConference of Engineers and Computer Scientists 2012, IMECS2012*, (Hong Kong, 2012).

THE SUBSYSTEM GROUPING SCHEME USING USE CASE DEPENDENCY GRAPH AND DOMAIN-SPECIFIC SEMANTIC MODEL FOR LARGE COMPLEX SYSTEMS

NANCHAYA KHRUEAHONG

Department of Computer Engineering, Faculty of Engineering
Chulalongkorn University, Bangkok Thailand E-mail: nanchaya.k@student.chula.ac.th

WIWAT VATANAWOOD

Department of Computer Engineering, Faculty of Engineering
Chulalongkorn University, Bangkok Thailand E-mail: wiwat@chula.ac.th

When a software system becomes large and complex, the UML use case diagram used to capture its requirements is consequently difficult and complicated. According to the huge numbers of UML elements found in the initial use case diagram, the subsystem grouping could be used to alleviate the complexity. The subsystem grouping scheme using the use case dependency graphs has been introduced as an alternative technique to identify the structural cohesion of the use cases and their relations. In our approach, the prerequisite preparation of the initial use case diagram is needed to ensure the well-formedness style and proper naming convention beforehand. Moreover, the refinement of the subsystem grouping is additionally proposed using the domain-specific semantic model. The final version of the resulting use case diagram with the relevant subsystems has been refined to ensure the improvement of the readability, understandability and context-relativity.

1. Introduction

The UML use case diagram is practically used by system analyst for object-oriented modeling in the very first draft of the target system [1], [2], [3], [4]. It is a visual model that captures the behavioral requirements of the system [5]. Unfortunately, the current software system becomes larger and more complex system. That causes the resulting huge and complex figure of use case diagram and it seems increasingly difficult to read and understand.

Information hiding technique and subsystem design are focused as the key issues in this paper. They help the system analyst organize and simplify the huge numbers of the UML elements within the use case diagram [6], [7]. However, the competency on how to specify a relevant subsystem is still not common for the analyst. It should be helpful if there is a guideline regarding these capabilities. We proposed an automatic subsystem grouping scheme, using use case dependency graph, to ease the drawing of the very first draft of a total use case diagram. The boundary of the subsystems would be recommended by the proposed scheme to increase the readability and understandability of the use case diagram in [8]. In this paper, we extend the grouping scheme to use the domain-specific semantic model as a thesaurus and then refine the subsystems accordingly to obtain the better context-related groups.

This paper is organized as follows. Section 2 reviews related works. Section 3 describes our proposed subsystem grouping scheme. In section 4, we demonstrate the case study. Finally, section 5 concludes the paper.

2. Related Works

To manage the complexity of requirements captured by the UML use case diagrams for large complex system is not well addressed in general. However, several best practices on how to draw the quality UML use case are concerned to alleviate the complexity. At the beginning, some templates and well-formedness rules [9] were formally defined, using set theory and logic, to ensure the syntactical constraints among use case elements and some guidelines are proposed to be followed. Moreover, the visualization and Aesthetics of the layout of use case diagrams apparently increases the readability and understandability, [10] proposed the deterministic layout algorithm to support drawing use case diagram nicely.

For the large complex system, some best practices using the top-down approach are still the effective ways to compromise with the complexity and the information coverage needed in the modeling. Some examples in [11], [12] show the evidences of how to use a hierarchical framework for use case diagramming of the large complex embedded systems.

3. Our Proposed Automatic Subsystem Grouping Scheme

3.1. *Preparation of the Initial Input Use Case Diagram*

In our approach, a raw input use case diagram, probably with a huge number of elements, should be initially prepared to conform to the well-formedness rules (WFR) of use case diagram [9] in order to ensure the syntactical consistence and completeness of the relations and their constraints among elements in the use case diagram. The WFRs include the techniques how to draw Actor, Use Case, Association, Generalization, <<include>> and <<extend>> relationship, etc.

With the given raw input as mentioned above, we would be ready to perform the subsystem grouping in two passes: Pass I - The subsystem grouping scheme using use case dependency graphs and Pass II -The refinement of the subsystem grouping using domain-specific semantic model. In our approach, we expect the improvement of the refinement process by consulting a context-related semantic model represented using SADL in [13] rather than enforcing the naming convention of use case names in the drawing in [8].

3.2. *The Well-Formedness Rules [9] Revisited*

According to [9], well-formedness rules are a set of syntactical constraints of UML elements and their relations, especially for the UML use case diagram. The following sentences show some examples of the WFRs written in the natural language:

- An actor must have a name and must be associated with at least one use case.
- Actors are not allowed to interact with other actors.
- A use case must have a name and every use case is involved with at least one actor.
- The <<include>> relationship links the source use case to the destination use case.

The rest of the WFRs are described in [9].

3.3. *The Semantic Modeling as a Thesaurus*

We build a semantic model that captures the relationship among the terms (also called as classes) found in the use case names. The generalization relationship is essential to our version of semantic model. Each relationship between two classes would be specified as a superclass-subclass relationship. If there exist subclass s1 and subclass s2 of the same superclass p1, then s1, s2 and p1 would be appearing in the same context boundary. However, we basically propose that the use cases which perform on subclass s1 and superclass p1 would be appropriately grouped into the same subsystem. Thus, both subclass s2 and superclass p1 would be grouped into the same system as well. Our semantic model is domain-specific and used to guide the subsystem grouping scheme in case of the similar terms or the context-related terms are considered.

As shown in figure 4, the use cases with term 'Course Information' and 'Registration Deadline' would be grouped into the same subsystem since the superclass-subclass relationship is obviously shown. The SADL [13] could be used as a tool to implement this semantic model.

3.4. Pass I: The Subsystem Grouping Scheme using Use Case Dependency Graphs [8]

The raw input use case diagram, which is prepared according to section 3.1, will be transformed into a set of use case dependency graphs and the first version of the resulting subsystems is then identified using Algorithm 1. The definitions and algorithm are shown as follow:

Definition 1: Use Case Dependency Graph, DG. A use case dependency graph is tuple *DG = (N, E)*. We define *N = ACTOR ∪ USECASE* and *E = ASSOC ∪ REL ∪GEN*. *ACTOR* is a set of actors and *USECASE* is a set of use cases in the diagram. *ASSOC* is a set of edges on *ACTOR x USECASE*, *REL* is a set of edges on *USECASE x USECASE*, and *GEN* is a set of edges on *USECASE x USECASE*. We also define *REL = {REL-INC} ∪ {REL-EXT} and GEN = {REL-GEN}* to cope with the type of relationships and generalization.

Definition 2: <<include>> relationship, REL-INC. An <<include>> relationship is 2-tuple *REL-INC = (baseUC, incUC)*, where *baseUC* is a set of the base use cases, and *incUC* is a set of the included use cases.

Definition 3: <<extend>> relationship, REL-EXT. An <<extend>> relationship is 2-tuple *REL-EXT = (baseUC, extUC)*, where *baseUC* is a set of the base use cases, and *extUC* is a set of the extending use cases. For <<extend>> relationship, we intentionally define the direction of the edge starting from the base use case to the extending use case.

Definition 4: Generalization relationship, REL-GEN. A generalization relationship is 2-tuple *GEN = (superUC, subUC)*, where *superUC* is a set of the parent use cases, and *subUC* is a set of the child or subordinate use cases. For generalization relationship, we intentionally define the direction of the edge starting from the parent use case to the child use case.

Algorithm 1: Subsystem Grouping
Input: A set of dependency graphs *TDG = {DG}* generated by definition 1-4

a)Dropout the DG_i which has number of nodes less than 3

   For each DG_i,

      If NumberOfNode(DG_i) < 3 then delete DG_i from TDG

b) Repeatedly find the subsystems

   Do while $TDG \neq \{\}$

      {
      Find the DG_i that has the maximum number of nodes and call it MaxDG

      For each DG_i
         {
         If the set of nodes of MaxDG $\cap$ the set of nodes of $DG_i \neq \{\}$ then

            o The new MaxDG equals MaxDG $\cup DG_i$

            o Delete DG_i

         }

      Define MaxDG as a subsystem.
      }

3.5. Pass II: The Refinement of the Subsystem Grouping using Domain-Specific Semantic Model

The intention of the refinement is to reconsider the dropout use cases during step a) of the algorithm 1 (Subsystem Grouping) and include them into the relevant subsystems. Our domain-specific semantic model is defined using the SADL's superclass-subclass relationships. As we mentioned earlier, the target subclass s1 and its immediate superclass p1 would be considered in the same subsystem.

180

4. The Case Study

This section demonstrates our subsystem grouping scheme by using a case study called "The course registration system". The raw input use case diagram of the course registration system is shown in figure 1. We follow the preparation steps in section 3.1 so that the use case diagram is now in the well-formedness style.

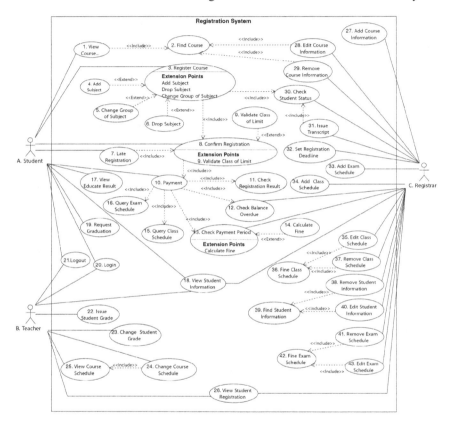

Figure 1. The Well-formedness Diagram of the Course Registration System

With the raw input use case diagram, the set of dependency graphs are defined and shown in figure 2. We found that 33 dependency graphs are generated. The graph number 2, 15, 9, 28, 30 and 32 are respectively selected as a MaxDG in step b) of algorithm 1 to form each subsystem. The graph number 5, 6, 7, 8, 9, 10, 11, 12, 13, 14, 16, 17, 20, 22, 23, 24, 25, 26 and 27 have been dropout and will be reconsidered in the refinement process.

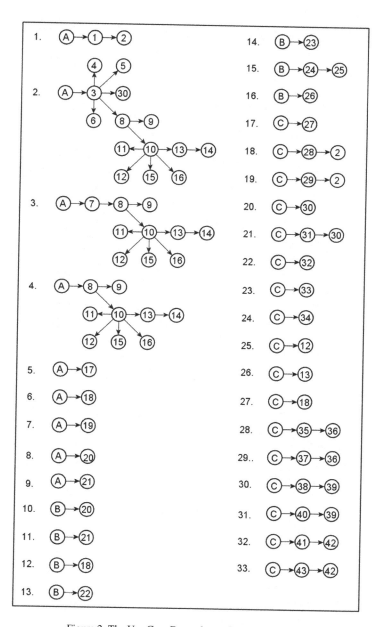

Figure 2. The Use Case Dependency Graphs Generated

182

As a result of the algorithm, six subsystems are identified and shown in figure 3. Each subsystem shows the appropriate use cases and their relationships.

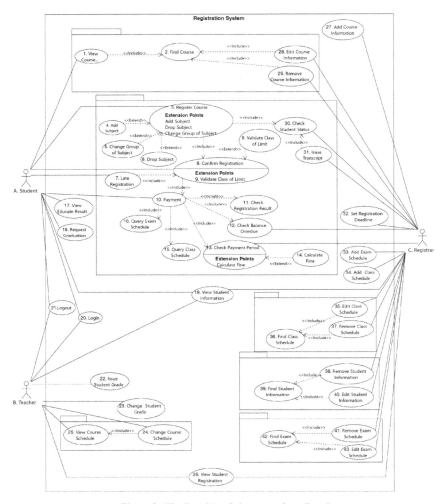

Figure 3. The Resulting Subsystems from Pass I

The figure 4 shows the part of the semantic model, the 'Registration Deadline' class is a subclass of 'Course Information'.

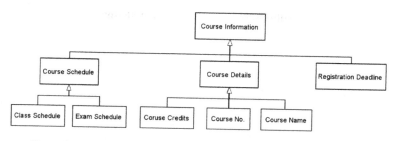

Figure 4. A Sample of Superclass-Subclass Relationships in Semantic Model

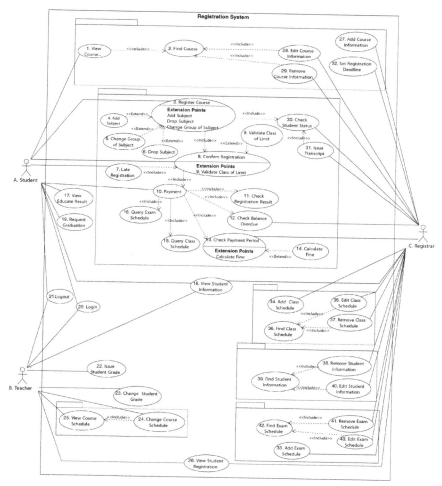

Figure 5. The Final Use Case Diagrams with the Relevant Subsystems.

In Figure 5, the resulting use case diagram with the relevant subsystem grouping is shown.

5. Conclusion

To read and understand the use case diagram for large complex system is a hard work for system analyst. The extension of subsystem grouping scheme using use case dependency graph is proposed. The subsystem grouping scheme consists of two passes of processing. We found that the initial use case diagram should be prepared in the systematic way. The well-formedness rules and the proper naming convention mentioned earlier are still recommended in the stage of preparation of the initial use case diagram.

Firstly, the use case dependency graphs are exploited to classify an initial set of subsystems in Pass I. Some unclassifiable use cases which have been dropout will be reconsidered in Pass II. The refinement of the subsystem grouping using semantic model is needed to ensure the completeness of the final result.

Practically, the metadata of the UML use case diagram would be represented in .XMI file format and the file could be automatically processed using our proposed scheme.

References

1. B. Dpbomg and E. Parson, "Dimensions of UML Diagram Use: A Survey of Practitioners", *Journal of Database Management 19(1)*, 2008, pp. 1-18.
2. G. Booch, J. Rumbaugh, and I. *Jacobson. The Unified Modeling Langueage User Guide*. USA: Addison-Wesley, 2001.
3. M. N. Alanazi, "Basic Rules to Build Correct UML Diagram," Proceeding of International Conference on New Trends in Information and Service Science (NISS'09), 2009, pp. 72-76.
4. Y. Labiche, "The UML Is More Than Boxes and Lines," Chaudron, M.R.V. (ed) Models in Software Engineering, *Springer-Verlag Berlin Heidelberg. LNCS*, Vol. 5421, 2009, pp. 375-386.
5. A. Cockburn, *Writing Effective Use Cases*. Addison-Wesley, 2001.
6. G.A. Kohring, "Complex Dependency in large Software Systems," *Journal of Advances in Complex Systems*, Vol. 12, No. 6, 2009, pp. 565-581.
7. C. R. Myers, "Software Systems as Complex Network: structure, function, and evolvability of software collaboration graphs," *Phys. Rev.* E68, 046116, 2003.
8. N. Khrueahong and W. Vatanawood, "An Automatic Subsystem Grouping using Use Case Dependency Graph for Large Complex Systems", *Proceedings of The International MultiConference of Engineers and*

Computer Scientists 2012, IMECS 2012, 14-16 March, 2012, Hong Kong, pp. 763-768.

9. N. Ibrahim, R. Ibrahim, M.Z. Saringat, D. Mansor, and T. Herawan, *"On Well-Formedness Rules for UML Use Case Diagram,"* Proceedings of the International Conference on Web Information System and Mining (WISM'10), *Springer-Verlag Berlin Heidelberg,* pp. 432-439, (2010).

10. H. Eichelberger, "Automatic Layout of UML Use Case Diagrams," Proceeding of the 4th ACM symposium on Software visualization (SoftVis'08), 2008.

11. K. S. Lew, T. S. Dillon, and K. E. Forward, "Software Complexity and Its Impact on Software Reliability," *IEEE Transaction on Software Engineering,* Vol. 14, No. 11, November, 1988.

12. E. Nasr, J. McDermin, and G. Bernat, "A Technique for managing Complexity of Use Cases for Large Complex Embedded Systems," *Proceedings of the Fifth IEEE International Symposium on Object-Oriented Real-Time Distributed Computing (ISOEC'02)* , 2002, pp. 225-232.

13. A. W. Crapo. SADL SoureForge Home [Online], Available: http://sadl.sourceforge.net.

MOBM: A METHODOLOGY FOR BUILDING PRACTICAL DOMAIN ONTOLOGIES FROM DATABASE INFORMATION

MINYOUNG RA, DONGHEE YOO, SUNGCHUN NO,
JINHEE SHIN, CHANGHEE HAN

Department of Electronics Engineering & Information Science,
Korea Military Academy, Seoul, Republic of Korea,
E-mail: {myra, dhyoo, is695, suhacci, chhan}@kma.ac.kr

Existing ontology development methodologies are divided into two approaches. One is to develop an ontology from database information; the other is to construct an ontology using domain terms according to a top-down method, bottom-up method, middle-out method, or hybrid method. To effectively represent organizational knowledge of an ontology, we have presented a mixed ontology building methodology (MOBM) that combines the characteristics of both approaches. The MOBM first creates a kernel ontology as the core using various types of database information, including database schema, and then completes the ontology by applying the top-down method and the bottom-up method, respectively.

1. Introduction

Recently, a growing need for research into effectively managing organizational knowledge using ontologies has emerged. In general, ontologies are formal and consensual specifications of conceptualizations that provide a shared understanding of a domain [1]. To increase the effect of knowledge management, the development of a well-defined ontology using various concepts of organizational knowledge is needed. However, the process of developing a suitable ontology for an organization's knowledge management involves significant time and considerable costs [2]. Accordingly, it is important to reuse pre-developed ontologies. However, it is difficult for an organization to identify a pre-developed ontology that expresses the organization's information appropriately; therefore, it is necessary to develop an ontology that is customized to the specific organization.

The core of ontology development is defining the key concepts necessary for the clear expression of an organization's knowledge and reducing the cost of development by simplifying the development process. One method is to use database information that is actually employed in the organization to the fullest.

A database is a depository of information, and in relational databases, the data are stored as tables. Such a database contains a large volume of information that is critical to the corresponding application domains. Thus, the use of well-organized information in a relational database allows for quicker and more accurate collection of the core concepts required for the development of a precise ontology for each task.

Existing ontology development methodologies are largely divided into two groups. The first methodology is created from the ontology of an existing database schema [3, 4, 5, 6, 7]. The second lies in the direction of conceptualizing ontology using the bottom-up method [8, 9], top-down method [10], middle-out method [11], or hybrid method [12]. However, many restrictions exist in terms of building an ontology that accurately expresses an organization's knowledge and information when using just one method. In the former method, the ontology expresses only the concepts in the database and, therefore, only those limited terms necessary for knowledge expression are included in the ontology. The methods in the latter group do not deal with database information that expresses an organization's knowledge.

In this paper, we present the mixed ontology building methodology (MOBM) that was proposed in our previous research [13]. The MOBM combines the characteristics of both approaches (existing database schema and conceptual-lizing ontology using various methods) to more effectively represent an organization's knowledge as ontology. The MOBM first creates a kernel ontology that is as extensive as possible; this then becomes the core that uses various types of database information, including database schema. The method completes the additional parts of the ontology by applying the top-down method and the bottom-up method, respectively. In addition, this paper describes the process for the staged application of the MOBM based on a scenario for a virtual database company and evaluates that methodology.

The paper is structured as follows. In Section 2, we review related works. In Section 3, we present an overview of the MOBM and describe the details of each step in this approach, and in Section 4, we introduce an exemplary scenario based on the MOBM. Finally, in Section 5 we draw conclusions, including lessons learned from this process, and suggest future research.

2. Related Works

Much research has been conducted on the issue of ontology building methodology. The related work can be divided into two categories. The first involves developing an ontology from database schemas. This work is generally undertaken from three directions: (1) Extract the ER model first from the database schema using reengineering, then extract the ontology from that model [3]; (2) given the database schema and ontology, for semantic web applications, extract mapping rules in common among them [4, 5]; and (3) generate the ontology structure itself from the relational database schema [6, 7].

The second category collects terminologies, and then builds the ontology by first analyzing concepts, forming a hierarchy for the concepts, and defining the relationships between the concepts and the rules for acquiring domain knowledge. According to the refinement process assigned to this task, the ontology is then completed. Several methods have been reported to accomplish this task. The bottom-up method starts from the most specific classes and then groups them into more general concepts [8, 9]. The top-down method starts with defining the most general concepts and then divides these general concepts into detailed sub-concepts [10]. The middle-out method starts with certain middle-level concepts and then applies the bottom-up method or the top-down method, as appropriate [11]. The hybrid method merges ontologies developed from the bottom-up method and top-down method, respectively, into one ontology [12].

In this paper, a mixed methodology is presented; the methodology first generates a kernel ontology that is as extensive as possible using database information and then completes the ontology by applying the bottom-up method and the top-down method, respectively, to build additional parts of the ontology.

3. The Mixed Ontology Building Methodology

3.1. *Overview of the MOBM*

An overview of the MOBM is depicted in Figure 1. In this methodology, mapping rules are defined to extract the main concepts and main relationships of a certain domain ontology from the target database schema. This kind of domain ontology is called the kernel ontology. The kernel ontology is enhanced by adding upper level terms and lower level terms, which are collected from domain knowledge or instances of the target database, because they may reflect new concepts or new relationships that did not exist in the target database schema. Based on the top-down method, the upper level terms are conceptualized into upper concepts. In the same way, the lower level terms are conceptualized into lower concepts using the bottom-up method. Once the upper concepts and lower concepts have been developed, they are then linked to the kernel ontology. The MOBM includes eight steps for building domain ontologies, as follows:

- Step 1. Extract kernel ontology from database schema.
- Step 2. Create class hierarchies from upper concepts.
- Step 3. Create class hierarchies from lower concepts.
- Step 4. Connect these class hierarchies to the kernel ontology.
- Step 5. Enhance the semantics between inter-terms.
- Step 6. Enhance any restrictions.
- Step 7. Enhance additional axioms and rules.
- Step 8. Complete the domain ontology.

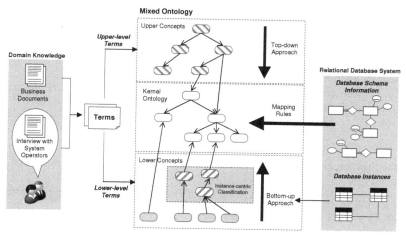

Figure 1. Overview of the MOBM.

3.2. *Extracting kernel ontology from database schema*

Many algorithms have been developed to extract a domain ontology from database information [3, 4, 5]. To achieve this, database schema has been used most frequently because it includes core concepts and relationships for properly building a domain ontology. In this step, we present a set of fundamental mapping rules to extract components of the kernel ontology from the database schema. Figure 2 lists the core database schema information used in the mapping rules.

> Database names, Relation names, Attribute names, Primary keys, Foreign keys, Attribute data types, M:N relationship constraints, Integrity constraints, Multi-valued attributes

Figure 2. Core information found in the database schema.

In this paper, kernel ontology is represented in Web Ontology Language (OWL). Thus, the mapping rules should include how to transfer the core information of the database schema into the components of OWL (e.g., *Class*, *Object property*, *Datatype property*). Thus, we used the following notations:

Suppose that the database schema (DS) has N tables. Then,

- T_i: the i-th table in DS where i = 1, 2, ..., N
- $Att_{i,j}$: the j-th attribute in T_i where j = 1, 2, ..., N_i
- PK_i: the set of the primary keys of T_i
- FK_i: the set of the foreign keys of T_i
- CK: the set of composite keys

Then, we arrive at the following equation formed from the definition:

$$DS = \bigcup_{i=1}^{N} T_i, \quad T_i = \bigcup_{j=1}^{N_i} Att_{i,j} \text{ for all } i$$

Extracting the kernel ontology from DS, we have

- C_k: the k-th class in the kernel ontology where k = 1, 2, ..., M $\leq$ N
- C: the set of classes in the kernel ontology, where

$$C = \bigcup_{k=1}^{M} C_k$$

Based on the notations developed above, the *mapping rules* are compiled as shown in Figure 3.

Rule 1: Find T_x for all x = 1, 2,..., N such that $T_x \notin C$. This rule has two cases:

(1) T_x such that $PK_x \subset CK$ and $T_x - PK_x = FK_x$: This case corresponds to the M:N relationship.

(2) T_x such that $PK_x \subset CK$, $\exists FK_x$, $FK_x \subset PK_x$ and $T_x - PK_x = \emptyset$: This case addresses the multi-valued attribute.

Rule 2: Map all other tables onto the ontology classes except the tables corresponding to Rule 1 above.

Rule 3: Specify the properties between the classes. For some y = 1, 2,..., N,

(1) if $\exists FK_y$ and $FK_y = PK_y$, then set up the subclass relationship between those two classes.

(2) if $\exists FK_y$ and $FK_y \neq PK_y$, then establish the referential integrity constraint between those two classes.

(3) $PK_y \subset CK$, $\exists FK_y$, $FK_y \subset PK_y$ and $T_y - PK_y \neq \emptyset$ (case of weak entity), then set up the is-part-of Object property between those two classes.

Rule 4: If the M:N relationship exists, set up the inverse Object property between those two classes.

Rule 5: For the case of the table, which treats the multi-valued attribute, T_z where z = 1, 2,..., N, PK_z ($\neq FK_z$) is identified as the Datatype property of the referencing class, and the maximum cardinality of the property has to be considered.

Rule 6: Specify the Datatype property for the remaining columns that are non-FK attributes of the table.

Figure 3. Mapping rules for extracting kernel ontology.

3.3. *Creating class hierarchies from upper concepts*

In this step, the upper concepts of the kernel ontology are conceptualized based on domain knowledge, such as an interview with the system operator or business documents. To this end, we first select new terms that do not exist in the target database schema from the domain knowledge. Among these selected terms, we identify upper level terms that can be defined in the upper concepts of the kernel ontology. These upper level terms are conceptualized into upper concepts using the top-down method.

3.4. *Creating class hierarchies from lower concepts*

This step specifies the lower concepts of the kernel ontology. The lower level terms are collected from the instances in the target database and the domain knowledge. Then, the bottom-up method is adopted to build the lower concepts. To accomplish this task, first the lower level terms are identified as the most specific individuals, and then we generalize them into more abstract concepts. Therefore, some instances of the database can be defined in a concept.

3.5. *Connecting these class hierarchies to the kernel ontology*

In this step, the upper concepts from section 3.3 and the lower concepts from section 3.4 are connected to the kernel ontology to integrate them into a single ontology. This ontology is called a mixed ontology. The connection methods are dynamically determined based on whether some concepts have the same name or not. The former case is automatically recognized in that it has the same concepts, and the latter case is semi-automatically recognized using machine learning methods, such as lexical checking and semantic checking.

3.6. *Enhancing the semantics between inter-terms*

The semantics between inter-terms can be obtained from the database schema or from domain knowledge. As mentioned earlier in section 3.2, some semantics are identified when the kernel ontology is extracted from the database information. Thus, this step defines the enhanced semantics between inter-terms that are not included in the database information. There are two types of enhanced semantics; one is related to class hierarchies and the other to the relationships between classes. In the former case, additional semantics for class hierarchies are defined not only as *subClassOf* but also as *equivalentClass*, *disjointWith*, and *intersectionOf*, among others. In the latter case, new relationships are specified among the upper concepts, lower concepts, and concepts of the kernel ontology when the mixed ontology is generated. In addition, additional semantics, such as *inverseOf, symmetric*, and *transitive*, can be defined in the mixed ontology when new semantics are discovered from the database information or domain knowledge.

3.7. *Enhancing any restrictions*

If restrictions are identified from the domain knowledge in this step, mixed ontology allows restrictions to be placed depending on how properties can be used by instances of a class. For example, we would say that a person has exactly one ID number, and that a seminar is presented by at least one presenter. Based on such restrictions, logical errors in facts and their relationships using the mixed ontology can be verified.

3.8. *Enhancing additional axioms and rules*

Either additional axioms or domain rules will be identified from the domain knowledge. To define formal concepts that are always true, a set of axioms (e.g., *subClassOf, equivalentClass, sameAs*) is used in the mixed ontology. A new class of mixed ontology can be built from existing components (class, properties, individual) by fitting them together into the definitions. In the case of domain rules, they are represented in the form of an '*If-Then*' structure. Domain experts can define various domain rules depending on the type of domain. These additional axioms or domain rules are then used to check logical correctness and infer additional knowledge through reasoning.

3.9. *Completing the domain ontology*

The final step is building a domain ontology in an ontology language, such as OWL. To do this, we used Protégé, which is an ontology building tool that implements the conceptual models designed in the previous steps into OWL.

4. Example Scenario for MOBM

In this section, we explain the process of constructing an ontology in a hypothetical scenario through applying the MOBM. The example company in our scenario was drawn up based on the well-known COMPANY database schema in the Elmasri/Navathe Book [14]. Figure 4 shows the relational database schema that would be extracted into the kernel ontology.

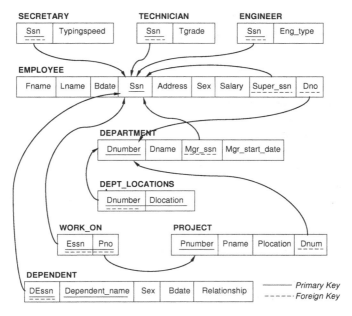

Figure 4. The relational database schema diagram [14].

Step 1: The following sub-steps summarize the process of extracting the kernel ontology from the database schema information in Figure 4 (this step corresponds to the mapping rules in Figure 3).

(1) Recognize the tables in the database schema that cannot be a class.

    1.1 Recognize WORKS_ON, which represents the M:N relationship.

    1.2 Recognize DEPT_LOCATIONS, which addresses a multi-valued attribute.

(2) Map the other seven tables, EMPLOYEE, DEPARTMENT, PROJECT, DEPENDENT, SECRETARY, TECHNICIAN, and ENGINEER onto each ontology class.

(3) Set up the properties between the recognized classes as follows:

    3.1 Set up the subclass-relationships.

       - EMPLOYEE-SECRETARY

       - EMPLOYEE-TECHNICIAN

       - EMPLOYEE-ENGINEER

    3.2 Set up the referential integrity Object properties and identify their domains and ranges for the ontology.

       - Super_ssn, Dno for EMPLOYEE

       - Mgrssn for DEPARTMENT

- Dnum for PROJECT, DEssn for DEPENDENT

For example, Object property Dno has EMPLOYEE as its domain and DEPARTMENT as its range.

   3.3 Set up the is_part_of the Object property between EMPLOYEE and DEPENDENT,

where DEPENDENT represents the weak entity.

(4) Set up the inverse Object property between EMPLOYEE and PROJECT that represents the M:N relationship.

(5) Identify Dlocation, which is not the FK of DEPT_LOCATIONS, as the Datatype property of DEPARTMENT.

(6) Identify the other non-FK attributes in each class as the Datatype properties of the ontology:

- Ssn, Fname, Lname, Bdate, Address, Sex, Salary of EMPLOYEE
- Dnumber, Dname, Mgr_start_date of DEPARTMENT
- Pnumber, Pname, Plocation of PROJECT
- Dependent_name, Sex, Bdate, Relationship of DEPENDENT
- Typingspeed of SECRETARY
- Tgrade of TECHNICIAN
- Eng_type of ENGINEER

Figure 5 describes the extracted kernel ontology based on the mapping rules.

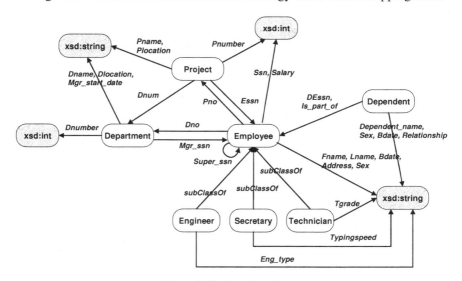

Figure 5. The kernel ontology.

Step 2: The terms that can be defined as the upper concepts of the kernel ontology are extracted from the domain knowledge. They are Company Entity, Task, People, Product, Owner, Project Member, and so forth. It is then verified that Company Entity includes the other terms from the extracted upper concepts. Hence, they are defined as subclasses of *Company Entity*. As we can see from Part a) in Figure 6, *People* is the subclass of *Company Entity*. In addition, *Employee* and *Dependent* in the kernel ontology, and the extracted terms *Owner* and *Project Member*, are defined as the subclasses of *People*. The subclasses of *People* are illustrated as a sibling relationship.

Step 3: After collecting the terms, defined as the lower concepts of the kernel ontology, from the domain knowledge and the instances in the database, we select only the terms that can be defined as a class. As you can see from Part b) in Figure 6, *Headquarters*, *Research,* and *Administration,* which are the attribute values of *Department,* each become a single class. This is because *accounting*, *marketing*, *finance*, and *human_resource*, which were collected from domain knowledge, are identified as the individuals of *Administration.*

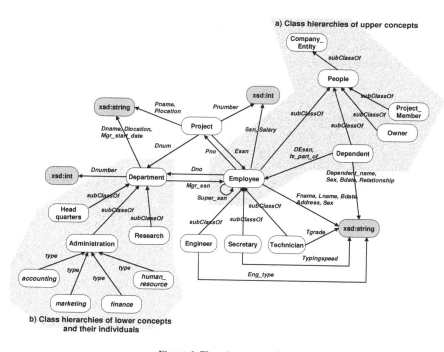

Figure 6. The mixed ontology.

Step 4: In this step, we construct a mixed ontology by linking the classes developed in Step 2 and Step 3. Figure 6 shows the parts of the mixed ontology built by connecting the upper ontology and the lower ontology to the kernel ontology.

Step 5: *Department*, *People*, *Product*, and *Task*, which are the subclasses of *Company Entity*, are set up to be *disjoint-relation* per the semantics of the class hierarchy. We also indicate that the individuals in each class cannot be the same because the individuals might be instances of all the classes owing to Open World Assumption. The next semantics, additional relationships, such as *Participant-relation* between class *Project Member* and class *Project*, are then added.

Step 6: The restrictions collected from the domain knowledge are added to the ontology. For example, the restriction saying that '*Department* has only and at least one *Mgrssn-relation* to *Employee*' can be expressed as follows:

$$Department \equiv \forall Mgrssn\ Employee \cap \exists Mgrssn\ Employee$$

Step 7: Additional axioms and rules for the mixed ontology are defined in this step. For instance, the axiom representing 'Project Member is People and has at least one Pno-relation to Project' can be expressed as follows:

$$Project\ Member \equiv People \cap \exists Pno\ Project$$

After defining axioms like this one, we are able to infer a new class hierarchy using the reasoner. In other words, *Project Member* can be inferred as a subclass of *Employee* because *Employee* is defined as the domain of *Pno*.

Step 8: We use Protégé to realize the mixed ontology defined by the MOBM in OWL form. Using Protégé, we can easily define Class, Object property, Datatype property, Restriction, Axiom, and so on, as shown in Figure 7 and Figure 8.

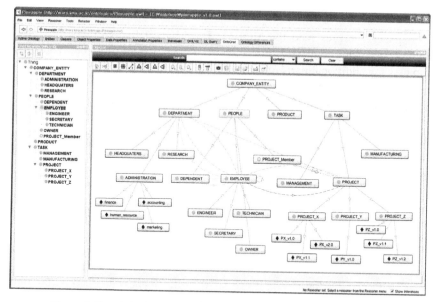

Figure 7. Completing class hierarchies of the mixed ontology using Protégé.

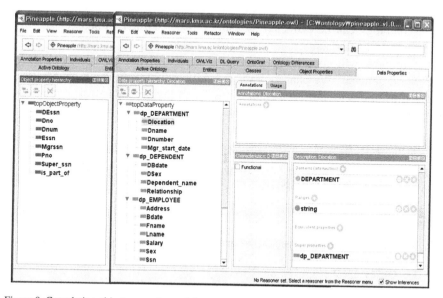

Figure 8. Completing object properties and datatype properties of the mixed ontology using Protégé.

5. Conclusion

In this paper, we presented the MOBM. The MOBM first extracts the kernel ontology by using database information to the extent possible and then completes the additional parts of the ontology by applying the top-down method and the bottom-up method, respectively. To show the application possibilities, we applied this methodology to the example company database and explained the process for building the ontology. Based on this application, the advantages of the MOBM are as follows: 1) Core classes needed to build an ontology can be clearly extracted; 2) object properties can be easily defined by using the table relationship in database schema, and datatype properties can also be easily defined by using the attribute information of tables in the database schema; and 3) the MOBM makes it possible to gather information quickly regarding the core concepts and relationships that ontologies comprise, so the time needed to build the ontology can be reduced.

However, the MOBM still has certain weaknesses: 1) If a domain has a small number of tables in the database schema, the number of classes that the kernel ontology can have is limited (in this case, the effectiveness attained by building the initial ontology would be reduced); 2) in addition, if a domain has a small number of 'subclass_of' relationships between tables, the hierarchy of the generated kernel ontology may be weak. In this case, more concepts for the kernel ontology can be extracted by using broader term (BT) and/or narrower term (NT) of the thesaurus.

Future work will continue in the direction of applying the methodology to real-world domains. We plan to develop a military ontology using the MOBM, apply it to the *Defense Information System*, and investigate and determine the degree to which this ontology is useful.

Acknowledgments

This research was supported by ADD (Agency for Defense Development). Contract Number UD110058MD.

References

1. T. Gruber, "A Translation Approach to Portable Ontology Specifications," *Knowledge Acquisition*, Vol. 5, No. 2, 1993, pp. 199-220.
2. D. Gašević, D. Djurić, V. Devedžić, "Model Driven Engineering and Ontology Development," Second Edition, Springer, Berlin, 2009.

3. J. Trinkunas, O. Vasilecas, "Building Ontologies from Relatinoal Databases Using Reverse Engineering Methods," *In Proceedings of International Conference on Computer Systems and Technologies*, 2007.
4. Z. Xu, S. Zhang, Y. Dong, "Mapping between Relational Database Schema and OWL Ontology for Deep Annotation," *In Proceedings of the 2006 IEEE/WIC/ACM International Conference on Web Intelligence*, 2006, pp. 548-552.
5. N. Konstantinou, D. E. Spanos, N. Mitrou, "Ontology and Database Mapping: A Survey of Current Implementations and Future Directions," *Journal of Web Engineering*, Vol. 7, No. 1, 2008, pp. 1-24.
6. N. Cullot, R. Ghawi, K. Yétongnon, "DB2OWL: A Tool for Automatic Database-to-Ontology Mapping," *In Proceedings of the 15th Italian Symposium on Advanced Database Systems (SEBD 2007)*, pp. 491-494.
7. S. S. Sane, A. Shirke, "Generating OWL Ontologies from a Relational Databases for the Semantic Web," In Proceedings of International Conference on Advances in Computing, Communication and Control, 2009, pp. 143-148.
8. M. Grüninger, M. S. Fox, "Methodology for the design and evaluation of ontologies," *In Proceedings of the Workshop on Basic Ontological Issues in Knowledge Sharing held in conjunction with IJCAI-95*, Montreal, Canada, 1995.
9. P. E. van der Vet, N. J. I. Mars, "Bottom-Up Construction of Ontologies," *IEEE Transactions on Knowledge and Data Engineering*, Vol. 10, No. 4, 1998, pp. 513-526.
10. G. Schreiber, B. Wielinga, W. Jansweijer, "The KACTUS view on the 'O' Word," *In IJCAI Workshop on Basic Ontological Issues in Knowledge Sharing, Montreal, Canada*, 1995, pp. 159-168.
11. O. Corcho, M. Fernández-López, A. Gómez-Pérez, A. López-Cima, "Building legal ontologies with METHONTOLOGY and WebODE," *In Proceedings of Law and the Semantic Web*, 2003, pp. 142-157.
12. F. J. Lopez-Pellicer, L. M. Vilches-Blázquez, J. Nogueras-Iso, O. Corcho, M. A. Bernabé, A. F. Rodríguez, "Using a hybrid approach for the development of an ontology in the hydrographical domain," *In Proceedings of 2nd Workshop Ontologies for Urban Development: Conceptual Models for Practitioners*, 2007.

13. M. Ra, D. Yoo, S. No, J. Shin, C. Han, "The Mixed Ontology Building Methodology Using Database Information," Lecture Notes in Engineering and Computer Science: Proceedings of The International MultiConference of Engineers and Computer Scientists 2012, IMECS 2012, 14-16 March, 2012, Hong Kong, pp. 68-73.
14. R. Elmasri, S. B. Navathe, "Fundamentals of Database Systems," Sixth Edition, Addison Wesley, 2010.

A TRIAL OF THE DYNAMIC WEIGHTED SUM METHOD FOR MULTI-OBJECTIVE OPTIMIZATION

HONG ZHANG

Department of Brain Science and Engineering, Kyushu Institute of Technology,
2-4 Hibikino, Wakamatsu, Kitakyushu 808-0196, Japan
E-mail: zhang@brain.kyutech.ac.jp

The key idea of the dynamic weighted sum (DWS) method is to systematically change the weights corresponding to each criterion for efficiently dealing with multi-objective optimization (MOO). In this paper, we investigate the search effect of the DWS method using the four kinds of dynamic weighted aggregations which are linear, bang-bang, sinusoidal, and random weighted ones, respectively. As a search optimizer, the method of cooperative PSO called multiple particle swarm optimizers with inertia weight (MPSOIWα) is used to find Pareto-optimal solutions. To demonstrate the search performance and effect, computer experiments on a suite of 2-objective optimization problems are carried out. The obtained performance results, i.e. the number, front distance, and cover rate of Pareto-optimal solutions corresponding to each given test problem, indicate that the linear weighted aggregation is the most suitable for acquiring search specifications.

1. Introduction

The process of optimizing simultaneously a collection of multiple objective functions subject to certain constraints is called multi-objective optimization (MOO) or multi-function optimization or multi-criterion optimization[4,6,26]. Since many practical problems are involved in MOO, which can be mainly found in different domains of science, technology, business, finance, automobile design, aeronautical engineering and so on[9,15,20,27], how to efficiently deal with MOO becomes a live issue, and is centered on the development, performance and effect of the treatment technique[12,21,36].

As to be well-known, there are lots of methods using fitness assignment strategies for MOO such as aggregation-based one, criterion-based one, and dominance-based one[8,16]. Among various manners, using the dynamic weighted sum (DWS) method which converts a multi-objective problem into a single objective one for finding Pareto-optimal solutions is a relatively simple multi-criterion decision analysis one.

The use of evolutionary computation (EC) such as genetic algorithms and genetic programming to MOO[28–30,35] has significantly grown in the last two decades[8,14]. As a new technique of swarm intelligence, in recent years, particle swarm optimization (PSO)[10,19] has been found to be successful extended to deal with MOO[18,34]. In comparison with the methods of EC, PSO seems particularly to be suitable for MOO for having good quality in convergence and solution accuracy[23,24,31]. For example, the methods of multiobjective particle swarm optimization (MOPSO)[7] and improved particle swarm optimizer with inertia weight (PSOIWα)[32] were proposed, and their search performance for handling MOO problems were analyzed.

Although good results can be obtained, their search abilities are inferior to that of the methods of cooperative PSO[2,3,13,22] search yet. For further improving the search performance of the PSOIWα, here, we propose to use a method of cooperative PSO called multiple particle swarm optimizers with inertia weight (MPSOIWα)[33] to search. The strong point of the MPSOIWα is to reinforce the search ability of the original PSOIWα by the union's power of plural swarms with hybrid search.

Due to the use of aggregation-based fitness assignment strategies, the evaluation criterion of the MPSOIWα itself successively varies corresponding to the weight combination. This matter means that the search environment of executing the DWS method is a dynamic one. Therefore, any optimization search method is required to quickly find Pareto-optimal solutions for accomplishing efficient search.

In order to investigate the characteristics of executing the DWS method, we use the four kinds of dynamic weighted aggregations which are linear, bang-bang, sinusoidal, and random weighted ones as known as well to examination. They provide numerous possibilities of efficiently finding Pareto-optimal solutions by changing the criterion with weight combination.

To demonstrate the effectiveness of the MPSOIWα run, computer experiments on a suite of 2-objective optimization problems are carried out. Based on the obtained experimental results, we indicate the character and effect of executing the DWS method corresponding to these weighted aggregations, respectively, and point out that which one of them is the most suitable for acquiring search specifications.

2. Basic Concepts

In this section, some concepts and definitions for dealing with MOO are introduced.

2.1. *MOO Problem*

Without loss of generality, a MOO problem can be described as follows:

$$Minimize_{\vec{x}} \quad \left(f_1(\vec{x}), f_2(\vec{x}), \cdots, f_I(\vec{x})\right)^T$$

$$s.t. \quad g_j(\vec{x}) \geq 0, \; j = 1, 2, \cdots, J \tag{1}$$
$$h_m(\vec{x}) = 0, \; m = 1, 2, \cdots, M$$
$$x_n \in [x_{nl}, x_{nu}], \; n \in (1, 2, \cdots, N)$$

where $f_i(\vec{x})$ is the i-th objective function, $g_j(\vec{x})$ is the j-th inequality constraint, $h_m(\vec{x})$ is the m-th equality constraint, $\vec{x} = (x_1, x_2, \cdots, x_N)^T \in \Re^N$ ($= \Omega$ decision variable space) is the vector of decision variable, x_{nl} and x_{nu} are the inferior and superior boundary values of each component x_n of the vector $\vec{x}$, respectively.

Because the I-objective functions ($I \geq 2$) may be conflicting with each other, there will be many solutions for a MOO problem. In this situation, a solution $\vec{x}^* \in \Omega$ is said to be a Pareto-optimal solution if and only if there does not exist another solution $\vec{x} \in \Omega$ so that $f_i(\vec{x})$ is dominated by $f_i(\vec{x}^*)$. The formula of the above relationship is given as

$$f_i(\vec{x}) \not\leq f_i(\vec{x}^*) \; \forall i \in I \quad iif \quad f_i(\vec{x}) \not< f_i(\vec{x}^*) \; \exists i \in I. \tag{2}$$

Furthermore, all of the Pareto-optimal solutions for a given MOO problem make up a Pareto-optimal solution set (P), or Pareto front (PF).

2.2. *Front Distance and Cover Rate*

For evaluating the performance results of handling MOO problems, both of the front distance (FD) and cover rate (CR) are used generally.

FD is a metric for checking how far the elements are in the set of non-dominated solutions found from those in the true Pareto-optimal solution set. It reflects estimation accuracy. The definition of FD is expressed as

$$FD = \frac{1}{Q}\sqrt{\sum_{q=1}^{Q} d_q^2}, \; d_q = f_i(\vec{x}_q^*) - f_i(\vec{x}_q^o), \; \forall i \in I \tag{3}$$

where Q is the number of the elements in the set of non-dominated solutions found, and d_q is the Euclidean distance (measured in objective space) between each of these obtained optimal solutions, $\vec{x}^o$, and the nearest member, $\vec{x}^*$, of the Pareto-optimal solution set.

CR is an other metric for checking the coverage of the elements being in the set of non-dominated solutions found to PF. This is because the estimation accuracy is insufficiency to reveal the distribution status of the obtained Pareto-optimal solutions and the possibility of necessary information to the decision-maker. The definition of CR is expressed as

$$CR = \frac{1}{I} \sum_{i=1}^{I} CR_i \tag{4}$$

where CR_i is the partial cover rate corresponding to the i-th objective, which is given by $CR_i = \sum_{l=1}^{\Gamma} \gamma_l / \Gamma$. Here, Γ is the number of dividing the i-th objective space which is from the minimum to the maximum of the fitness value, i.e. $[f_i(\vec{x})^{min}, f_i(\vec{x})^{max}]$, and $\gamma_l \in (0, 1)$ indicates the existence status of the obtained solutions in the l-th subdivision for the i-th objective.

2.3. Dynamic Weighted Sum Method

The popularity of using a weighted sum of objective functions is clear. However, a known drawback of the convenient weighted sum (CWS) method is that in case of a large number of objective functions, the appropriate weighting is difficult to choose a prior by the decision-maker.

To thoroughly conquer the weakness of the CWS method, the dynamic weighted sum (DWS) method is used to systematically deal with MOO in practice. The criterion F_d of the DWS method is expressed as

$$F_d(t, \vec{x}) = \sum_{i=1}^{I} c_i(t) f_i(\vec{x}) \tag{5}$$

where t is time-step to search, and $c_i(t) \geq 0$ is the dynamic weight satisfying the constraint $\sum_{i=1}^{I} c_i(t) = 1$.

For the sake of observation, a 2-objective optimization problem is considered to examination, and linear weighted aggregation (LWA), bang-bang weighted aggregation (BWA), sinusoidal weighted aggregation (SWA), and random weighted aggregation (RWA) are used in the DWS method for testing[17,32]. These dynamic weighted aggregations are defined as follows:

- LWA: $\quad c_1^l(t) = mod\left(\frac{t}{T}, 1\right), \qquad\qquad c_2^l(t) = 1 - c_1^l(t)$

- BWA: $\quad c_1^b(t) = \dfrac{sign\left(sin(2\pi t/T)\right)+1}{2}, \quad c_2^b(t) = 1 - c_1^b(t)$

- SWA: $\quad c_1^s(t) = \left|sin\left(\dfrac{\pi t}{T}\right)\right|, \qquad\qquad c_2^s(t) = 1 - c_1^s(t)$

- RWA: $\quad c_1^r(t) = \left.\dfrac{rand(t)}{t}\right|_{t\neq 0}, \qquad\qquad c_2^r(t) = 1 - c_1^r(t)$

where T is a period of the variable weights.

3. Algorithms

In this section, the algorithms of the PSOIW, PSOIWα, and MPSOIWα are described. For the sake of convenience to the following description, let the search space be N-dimensional, the number of particles in a swarm be L, the position of the i-th particle be $\vec{x}^i = (x_1^i, x_2^i, \cdots, x_N^i)^T \in \Omega$, and its velocity be $\vec{v}^i = (v_1^i, v_2^i, \cdots, v_N^i)^T \in \Omega$, respectively.

3.1. *PSOIW*

To overcome the weakness of the original PSO[1,5] in convergence, Shi et al. proposed to modify the update rule of the i-th particle's velocity by constant reduction of the inertia coefficient over time-step, and created the particle swarm optimizer with inertia weight (PSOIW)[11,25]. The mechanism of the PSOIW is expressed as

$$\begin{cases} \vec{x}_{k+1}^i = \vec{x}_k^i + \vec{v}_{k+1}^i \\ \vec{v}_{k+1}^i = w(k)\, \vec{v}_k^i + w_1 \vec{r}_1 \otimes (\vec{p}_k^i - \vec{x}_k^i) + w_2 \vec{r}_2 \otimes (\vec{q}_k - \vec{x}_k^i) \end{cases} \tag{6}$$

where w_1 and w_2 are coefficients for individual confidence and swarm confidence, respectively. $\vec{r}_1, \vec{r}_2 \in \Re^N$ are two random vectors, each element of which is uniformly distributed on the interval $[0, 1]$, and the symbol $\otimes$ is an element-wise operator for vector multiplication. $\vec{p}_k^i \; (= arg \max_{k=1,2,\cdots} \{g(\vec{x}_k^i)\}$, where $g(\vec{x}_k^i)$ is the criterion value of the i-th particle at time-step k) is the local best position of the i-th particle up to now, $\vec{q}_k (= arg \max_{i=1,2,\cdots} \{g(\vec{p}_k^i)\})$ is the global best position among the whole particles at time-step k. $w(k)$ is the variable inertia weight as follows:

$$w(k) = w_s + \frac{w_e - w_s}{K} \times k \tag{7}$$

where w_s and w_e are starting and terminal values of the variable inertia weight, respectively. K is the number of iteration for the PSOIW run. In the original PSOIW, w_s=0.9, w_e=0.4, and $w_1 = w_2 = 2.0$ are used.

3.2. *PSOIWα*

Although the convergence of the PSOIW is massively improved by the inertia weight strategy, but it still has a shortcoming in optimization, i.e. easily to fall into a local minimum and hardly to escape from that solution specially for handling multimodal problems.

For alleviating the weakness of the PSOIW search, we introduce the localized random search (LRS) into the PSOIW to make up a hybrid search

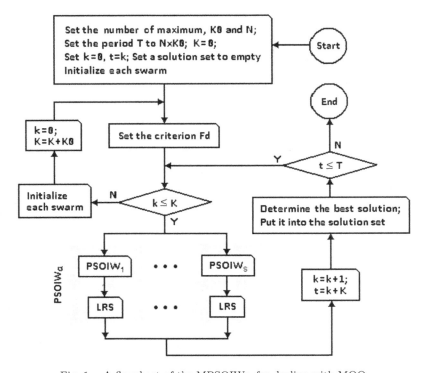

Fig. 1.   A flowchart of the MPSOIWα for dealing with MOO.

(called PSOIWα)[32]. It realizes to correct the solution found by the PSOIW in the search process. Namely, the best solution (position) of the PSOIWα search at time-step k is determined by

$$\vec{q}_{k+1} = \begin{cases} \arg \max\limits_{u=1,2,\cdots} \{g(\vec{q}_k + \vec{z}_u)\}, & if \ g(\vec{q}_k + \vec{z}_u) \geq g(\vec{q}_k) \\ \vec{q}_k, & otherwise \end{cases} \quad (8)$$

where $\vec{z}_u \in \Re^N \sim N(0, \sigma^2)$ is the u-th random data.

3.3. MPSOIWα

For further improving the search ability of the PSOIWα, we propose to use the method of multiple particle swarm optimizers with inertial weight (MPSOIWα) to search. Fig. 1 gives a flowchart of the MPSOIWα run for showing the whole treatment processing and information control by executing the DWS method.

We can see from Fig. 1 that the most difference between the PSOIWα ($S = 1$) and MPSOIWα ($S \geq 2$) is just to implement the plural PSOIWα in parallel. Due to the parallel processing, theoretically, the relation of the inequality $\max\limits_{s=1,\cdots,S}\{g(\vec{q}_k^s)\} \geq g(\vec{q}_k^s)$ is true. Hence, the probability of discovering good solutions becomes big. As a consequence, the search performance of the MPSOIWα is superior to that of the PSOIWα to finally achieve the improvement of search ability.

Thereafter, from the all of the solutions obtained by each PSOIWα at time-step k, the best solution of the multi-swarm hybrid search is determined by the maximum selection, $\vec{x}_k^o = arg \max\limits_{s=1,\cdots,S}\{g(\vec{q}_k^s)\}$. And then put the solution $\vec{x}_k^o$ into a solution set which is the storage memory of the multi-swarm.

In addition, it is to be noted that if the LRS is not implemented after the PSOIW run, the method will be called as MPSOIW.

4. Computer Experiments

In this section, the experimental setting is illustrated. To facilitate comparison and analysis of the search performance of the MPOSIWα, the suite of 2-objective optimization problems[36] in Table 1 is used in the following computer experiments. The characteristics of Pareto fronts of the given test problems include the convex (*ZDT1* problem), concave (*ZDT2* problem), and discontinuous multi-convex (*ZDT3* problem).

Table 1.   A suite of 2-objective optimization problems.

problem	2-objective function	search range
ZDT1	$f_{11}(\vec{x}) = x_1,$ $f_{12}(\vec{x}) = g(\vec{x})\left(1 - \sqrt{\dfrac{f_{11}(\vec{x})}{g(\vec{x})}}\right)$, where $g(\vec{x}) = 1 + \dfrac{9}{N-1}\sum\limits_{n=2}^{N} x_n$	$\Omega \in [0,1]^N$
ZDT2	$f_{21}(\vec{x}) = x_1,$ $f_{22}(\vec{x}) = g(\vec{x})\left(1 - \left(\dfrac{f_{21}(\vec{x})}{g(\vec{x})}\right)^2\right)$	$\Omega \in [0,1]^N$
ZDT3	$f_{31}(\vec{x}) = x_1,$ $f_{32}(\vec{x}) = g(\vec{x})\left(1 - \sqrt{\dfrac{f_{31}(\vec{x})}{g(\vec{x})}} - \left(\dfrac{f_{31}(\vec{x})}{g(\vec{x})}\right)sin\left(10\pi f_{31}(\vec{x})\right)\right)$	$\Omega \in [0,1]^N$

Table 2 gives the major parameters of the MPSOIWα run for handling the given 2-objectives optimization problems in Table 1.

Table 2. Major parameters of the MPSOIWα run.

parameter	value	parameter	value
the number of particles, L	10	the number of iterations, K	25
the number of period, T	2500	the number of random points, U	10
the search range of the LRS, σ	0.1	the number of multiple swarms, S	3

4.1. Performance Comparison

Fig. 2 shows the resulting solution distributions of the MPSOIWα and MPSOIW by respectively using the LWA, BWA, SWA, and RWA to handle each test problem. For quantitative analysis to the experimental results of the two search methods, Table 3 gives the statistical data, i.e. the obtained number of the solutions, and the corresponding FD and CR.

Table 3. Performance comparison of both the MPSOIWα and MPSOIW by using the LWA, BWA, SWA, and RWA, respectively (Γ is set to 100).

problem	agg.	MPSOIWα			MPSOIW		
		solu.	FD	CR (%)	solu.	FD	CR (%)
ZDT1	LWA	1254	2.234×10^{-8}	99.5	1191	3.948×10^{-8}	99.5
	BWA	187	9.809×10^{-5}	52.0	227	1.107×10^{-4}	53.0
	SWA	988	4.511×10^{-8}	99.5	1016	7.355×10^{-8}	99.0
	RWA	1383	3.869×10^{-8}	99.5	1410	1.221×10^{-7}	99.0
ZDT2	LWA	272	1.198×10^{-8}	94.0	283	1.992×10^{-7}	94.0
	BWA	259	3.692×10^{-7}	92.0	228	8.852×10^{-7}	91.5
	SWA	229	7.604×10^{-8}	93.5	219	3.381×10^{-7}	93.0
	RWA	256	4.636×10^{-8}	93.0	314	2.517×10^{-7}	93.0
ZDT3	LWA	1231	8.961×10^{-7}	46.0	1107	9.245×10^{-7}	45.5
	BWA	396	1.655×10^{-4}	40.5	421	6.551×10^{-5}	40.0
	SWA	949	9.433×10^{-7}	42.5	1018	1.092×10^{-6}	42.0
	RWA	1289	9.426×10^{-7}	43.5	1306	3.142×10^{-6}	41.5

Note: The values with underbar signify the best result for each given problem.

By comparing each performance index in Table 3, the following character and judgment can be obtained: Firstly, there is the most number of solutions obtained by using the RWA for the given test problems even for the ZDT2 problem in where a large number of Pareto-optimal solutions are

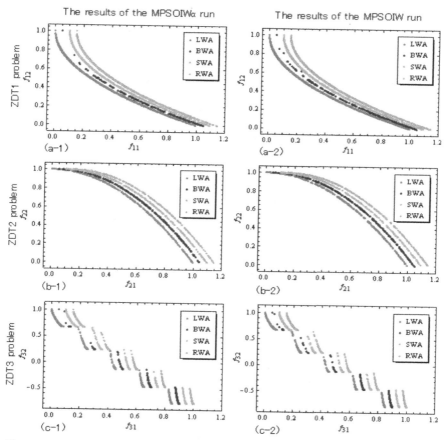

Fig. 2. Solution distributions of the MPSOIWα and MPSOIW by using the LWA, BWA, SWA, and RWA, respectively. Note: the distance between the experimental data sets for each subgraph is 0.05 (shift only in horizontal direction).

in unstable position[17]; Secondly, the solution accuracy of the MPSOIWα is superior to that of the MPSOIW for each test problem; Thirdly, the obtained results of using the LWA in CR index are the best than that of using BWA, SWA, and RWA, respectively; Fourthly, the search performance of using the LWA is not only much better than that of using the BWA, but also is relatively better than that of using the SWA or RWA as a valid conclusion in handling the given 2-objective optimization problems.

Owing to the above mentioned analysis, the effectiveness and search ability of the MPSOIWα are roughly confirmed. Furthermore, better solution distribution and higher solution accuracy can be observed as well. These results indicate that the fine change of the weight values in the DWS method can make that the probability finding good solutions greatly goes up as evidence. Based on the performance comparison of using the LWA, BWA, SWA, and RWA, the relationship of domination reflecting the search performance (SP) of the MPSOIWα by using each weighted aggregation can be given as follows:

$$LWA_{SP} \succeq RWA_{SP} \succ SWA_{SP} \succ BWA_{SP}$$

4.2. *Effect of Multi-swarm Search*

For further verifying the performance characteristics of the MPSOIWα, the following computer experiments on comparing the search performance of multi-swarm search and single swarm search with the different number of the particles used in the PSOIWα are carried out.

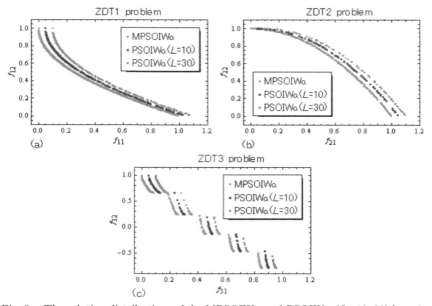

Fig. 3. The solution distributions of the MPSOIWα and PSOIWα (L=10, 30) by using the LWA. Note: the distance between the experimental data sets for each subgraph is 0.05 (shift only in horizontal direction).

As an example, Fig. 3 shows the resulting solution distributions of both the MPSOIWα and PSOIWα ($L = 10, 30$) by using the LWA. We can clearly see that the density of solution distributions of the MPSOIWα are higher than that of the PSOIWα run for each test problem under the condition of same number of period, $T = 2500$.

Table 4 gives the statistical data concerning each solution distribution shown in Fig. 3. By comparing the statistical results of the MPSOIWα and PSOIWα, big difference between the both experimental results can be observed. It clearly indicates the strong point of the MPSOIWα in search ability, which is not only to efficiently find a large number of *Pareto*-optimal solutions, but also to obtain them with high-accuracy.

Table 4. Performance comparison between the MPSOIWα and PSOIWα ($L = 10, 30$) by using the LWA (Γ is set to 100).

problem	method	solution	FD	CR (%)
ZDT1	MPSOIWα	1254	2.234×10^{-8}	99.5
	PSOIWα($L = 10$)	422	3.704×10^{-7}	89.5
	PSOIWα($L = 30$)	522	6.661×10^{-8}	91.0
ZDT2	MPSOIWα	272	1.198×10^{-8}	94.0
	PSOIWα($L = 10$)	102	4.338×10^{-8}	60.0
	PSOIWα($L = 30$)	231	9.938×10^{-8}	61.5
ZDT3	MPSOIWα	1231	8.961×10^{-7}	46.0
	PSOIWα(P=10)	423	6.748×10^{-6}	45.0
	PSOIWα(P=30)	432	4.496×10^{-6}	41.0

Note: The values with underbar signify the best result for each given problem.

On the other hand, The results in Table 4 show that the increment of the number of particles can cause the improvement of performance in the number of solutions and solution accuracy, but not in the cover rate for the *ZDT3* problem. It reflects the basic feature of the PSOIWα run to deal with MOO, i.e. the increment of particles used is not in proportion to the increment of *CR*.

5. Conclusions

As a simple method of cooperative PSO applied to MOO, in this paper, multiple particle swarm optimizers with inertia weight (MPSOIWα) has been presented. Applications of the MPSOIWα to the given suite of 2-objective optimization problems well demonstrated its effectiveness by executing the dynamic weighted sum (DWS) method.

From the resulting experimental data, it is observed that the search performance of the MPSOIWα is superior to that of the MPSOIW and PSOIW, and the search performance and effect of using the LWA is better than that of using the BWA or SWA or RWA. These results clearly indicate that plural swarm search and hybrid search are very effective to handle the given test problems by executing the DWS method. Therefore, it is no exaggeration to say that the obtained experimental results could offer an important evidence, i.e. choosing the DWS method by using the LWA or even RWA for efficiently handling complex MOO problems.

According to the above analysis, the significance of using the method of multi-swarm hybrid search to deal with MOO is notably effective. Accordingly, it can be said that a trend toward growth of cooperative PSO technique for improving the search performance is right.

Acknowledgment

This research was supported by Grant-in-Aid Scientific Research(C) (22500132) from the Ministry of Education, Culture, Sports, Science and Technology, Japan.

References

1. van den Bergh, F. (2002). *An Analysis of Particle Swarm Optimizers*, Ph.D thesis, University of Pretoria, Pretoria, South Africa.
2. van den Bergh, F. and Engelbrecht, A. P. (2004). *A cooperative approach to particle swarm optimization, J. IEEE Transactions on Evolutionary Computation*, **8**, 3, pp.225-239.
3. Chang, J. F., Chu, S. C., Roddick, J. F. and Pan, J. S. (2005). *A parallel particle swarm optimization algorithm with communication strategies, J. Information Science and Engineering*, 21, pp.809-818.
4. Chinchuluun, A. and Pardalos, P. M. (2007). *A survey of recent developments in multiobjective optimization, J. Ann Oper Res*, 154, pp.29-50.
5. Clerc, M. and Kennedy, J. (2000). *The particle swarm-explosion, stability, and convergence in a multidimensional complex space, J. IEEE Transactions on Evolutionary Computation*, **6**, 1, pp.58-73.
6. Coello Coello, C. A. and Lechuga, M. S. (2002). MOPSO: A proposal for multiple objective particle swarm optimization, in *Proc. Congress Evolutionary Computation (CECf2002)* (Honolulu, HI, USA), vol.1, pp.1051-1056.
7. Coello Coello, C. A., Pulido, G. T. and Lechuga, M. S. (2004). *Handling Multiple Objectives With Particle Swarm Optimization, IEEE Transaction on Evolutionary Computation*, **8**, 3, pp.256-279.
8. Deb, K. (2001). *Multi-Objective Optimization using Evolutionary Algorithms*, John Wiley & Sons, New York.

9. Deb, K. (2005). Multi-Objective Optimization, in E. K. Burke et al. (eds.), *Search Methodologies — Introductory Tutorials in Optimization and Decision Support Techniques*, Springer, pp.273-316.

10. Eberhart, R. C. and Kennedy, J. (1995). A new optimizer using particle swarm theory, in *Proc. the sixth Int. Symposium on Micro Machine and Human Science* (Nagoya, Japan), pp.39-43.

11. Eberhart, R. C. and Shi, Y. (2000). Comparing inertia weights and constriction factors in particleswarm optimization, in *Proc. 2000 IEEE Congress on Evolutionary Computation* (La Jolla, CA, USA), vol.1, pp.84-88.

12. Eghbal, M., Yorino, N., Zoka, Y. and El-Araby, E. E. (2009). *Application of Multi-Objective Evolutionary Optimization Algorithms to Reactive Power Planning Problem, IEEJ Transactions on Electrical and Electronic Engineering*, 4, pp.625-632.

13. El-Abd, M. and Kamel, M. S. (2008). *A Taxonomy of Cooperative Particle Swarm Optimizers, Int. J. Computational Intelligence Research*. **4**, 2, pp.137-144.

14. Hajela, P. and Lin, C.-Y. (1992). *Genetic search strategies in multicriterion optimal design, J. Operational Research*, 4, pp.99-107.

15. Hema, C. R., Paulraj, M. P., Nagarajan, R., Yaacob, S. and Adom, A. H. (2008). *Application of Particle Swarm Optimization for EEG Signal Classification, J. Biomedical Soft Computing and Human Sciences*, **13**, 1, pp.79-84.

16. Hughes, E. J. (2008). *Multiobjective Problem Solving from Nature, J. Natural Computing Series*, Part IV, pp.307-329.

17. Jin, Y., Olhofer, M. and Sendhoff, B. (2001). Dynamic Weighted Aggregation for Evolutionary Multi-Objective Optimization: Why Does It Work and How?, in *Proc. Genetic and Evolutionary Computation Conference (GECCO2001)* (San Francisco, CA, USA), pp.1042-1049.

18. Jin, Y. and Sendhoff, B. (2004). *Constructing Dynamic Optimization Test Problems Using the Multi-Objective Optimization Concept*, in G. Raidl et al. (eds.) *Applications of Evolutionary Algorithms*, LNCS 3005, pp.525-536.

19. Kennedy, J. and Eberhart, R. C. (1995). Particle swarm optimization, in *Proc. 1995 IEEE Int. Conference on Neural Networks* (Piscataway, NJ, USA), pp.1942-1948.

20. Li, X., Branke, J. and Kirley, M. (2007). On Performance Metrics and Particle Swarm Methods for Dynamic Multiobjective Optimization Problems, in *Proc. IEEE Congress of Evolutionary Computation (CEC)* (Singapore), pp.1635-1643.

21. Marler, R. T. and Arora, J. S. (2004). *Survey of multi-objective optimization methods for engineering, J. Struct Multidisc Optim*, 26, pp.369-395.

22. Niu, B., Zhu, Y. and He, X. (2005). Multi-population Cooperation Particle Swarm Optimization, in M. Capcarrere et al. (eds.), *Advances in Artificial Life*, LNCS 3630 (Springer Heidelberg), pp.874-883.

23. Poli, R., Kennedy, J. and Blackwell, T. (2007). *Particle swarm optimization — An overview, J. Swarm Intell*, 1, pp.33-57.

24. Reyes-Sierra, M. and Coello Coello, C. A. (2006). *Multi-Objective Particle Swarm Optimizers: A Survey of the State-of-the-Art, Int. J. Computational*

Intelligence Research, **2**, 3, pp.287-308.

25. Shi, Y. and Eberhart, R. C. (1998). A modified particle swarm optimiser, in *Proc. IEEE International Conference on Evolutionary Computation* (Anchorage, Alaska, USA), pp.69-73.

26. Steuer, R. E. (1986). *Multiple Criteria Optimization: Theory, Computations, and Application*, John Wiley & Sons, New York.

27. Suzuki, S. (1997). *Application of Multi-objective Optimization and Game Theory to Aircraft Flight Control Problem*, J. Institute of Systems, Control and Information Engineers, **41**, 12, pp.508-513 (in Japanese).

28. Tripathi, P. K., Bandyopadhyay, S. and Pal, S. K. (2007). *Multi-Objective Particle Swarm Optimization with time variant inertia and acceleration coefficient*, J. Information Sciences, 177, pp.5033-5049.

29. Tsou, C.-S., Chang, S.-C. and Lai, P.-W. (2007). *Using Crowing Distance to Improve Multi-Objective PSO with Local Search*, in Felix T. S. Chan et al. (eds.), *Swarm Intelligence: Focus on Ant and Particle Swarm Optimization*, pp.77-86.

30. van Veldhuizen, D. A. and Lamont, G. B. (1998). Multiobjective evolutionary algorithm research: A history and analysis, Dept. Elec. Comput. Eng., Graduate School of Eng., Air Force Inst. Technol., Wright-Patterson AFB, OH, Tech. Rep. TR-98-03.

31. Zhang, H. (2011). Assessment of An Evolutionary Particle Swarm Optimizer with Inertia Weight, in *Proc. 2011 IEEE Congress on Evolutionary Computation (CEC)* (New Orleans, USA), pp.1746-1753.

32. Zhang, H. (2012). *The Performance Analysis of an Improved PSOIW for Multi-objective Optimization, IAENG Int. J. Computer Scientists*, **39**, 1, pp.34-41.

33. Zhang, H. (2012). Multiple Particle Swarm Optimizers with Inertia Weight for Multi-objective Optimization, in *Lecture Notes in Engineering and Computer Science: Proc. Int. MultiConference of Engineers and Computer Scientists 2012 (IMECS 2012)* (Hong Kong), pp.23-28.

34. Zhang, H. (2012). *An Analysis of Multiple Particle Swarm Optimizers with Inertia Weight for Multi-objective Optimization, IAENG Int. J. Computer Scientists*, **39**, 2, pp.190-199.

35. Zhou, A., Jin, Y., Zhang, Q., Sendhoff, B. and Tsang, E. (2007). Prediction-based Population Re-initialization for Evolutionary Dynamic Multi-objective Optimization, in *Proc. 4th Int. Conf. on Evolutionary Multi-Criterion Optimization* (Matsushima, Japan), pp.832-846.

36. Zitzler, E., Deb, K. and Thiele, L. (2000). *Comparison of Multiobjective Evolutionary Algorithms: Empirical Results, J. Evolutionary Computation*, **8**, 2, pp.173-195.

A MULTI-AGENT PLATFORM TO MANAGE DISTRIBUTED AND HETEROGENEOUS KNOWLEDGE BY USING SEMANTIC WEB

INAYA LAHOUD

*SET Laboratory, University of Technology of Belfort-Montbeliard
Belfort, 90000, France, Inaya.Lahoud@utbm.fr*

DAVY MONTICOLO

*ERPI Laboratory, Polytechnic National Institute of Lorraine
Nancy, 54000, France, Davy.Monticolo@ensgsi.inpl-nancy.fr*

VINCENT HILAIRE

*SET Laboratory, University of Technology of Belfort-Montbeliard
Belfort, 90000, France, Vincent.Hilaire@utbm.fr*

SAMUEL GOMES

*M3M Laboratory, University of Technology of Belfort-Montbeliard
Belfort, 90000, France, Samuel.Gomes@utbm.fr*

Nowadays new product development involves different types of actors (technician, managers, and board of directors) which must be able to share knowledge, experiences and work together efficiently. Each actor has a professional specialty and uses one or several software tools (CAO, project management tools, PLM tools ...) dedicated to her specific activities. Each of these software tools produces different information sources (databases, XML files, text files) which are distributed through the enterprise network. In this paper, we present the design of a multi-agent software architecture that allows the capitalization of distributed and heterogeneous knowledge. We then describe how the agents handle knowledge through ontologies and build semantic queries.

1. Introduction

Current competitiveness has led companies to a fast product renewal coupled with lower costs. Today, manufacturers are creating more and more efficient products while meeting shorter deadlines in order to satisfy customer needs and sales.

These companies have to break into new markets by showing how creative they are to grow more profitable. Such creativity requires optimized organization mastery, a control of the industrial process and the development of a 'learning company' in which getting knowledge benefits ongoing projects. Learning within the company has now become the best way to be competitive. Learning is not only about improving, but also a way to start a new 'learning culture' in which every co-worker, every team and the whole company will be able to optimize their capacities by sharing their knowledge and their know-how. Thus the industrial interest in methodologies and tools enabling capitalization and management of distributed and heterogeneous knowledge has been growing stronger. This paper describes the design of a MAS platform that shall manage knowledge coming from different information sources. The first part briefly describes the problematic of the management of distributed and heterogeneous knowledge. The second part will explain the design rationale of the SemKnow [13] MAS platform. The last section focuses on our current work on how the agents can manage the problem of knowledge extraction from different software tool databases by using domain ontologies.

2. The Agent Paradigm Used in Knowledge Engineering

2.1. *MAS Used in Knowledge Engineering*

The aim of knowledge engineering is to gather, study, organize and represent knowledge. Multi-agent systems have already proved their efficiency to support such tasks. Klusch made a list of the services that a multi-agent system can offer in a knowledge management approach [11]:

- search, acquire, analyse and classify knowledge coming from various information sources;
- Give information to human and computing networks once usable knowledge is ready to be consulted;
- Negotiate on knowledge integration or exclusion into the system;
- Give explanation to the quality and reliability related to the integrated knowledge;
- Learn progressively all along the knowledge management process;

Such services are mostly implemented to create two MAS categories devoted to knowledge management. The first MAS type is based upon an agent cooperation to solve complicated problems related to knowledge types. Some of these SMAs were created as complementary tools in information management (workflow, ontologies, information research systems and so on) to design platforms like FRODO [1], CoMMA [8], Edamok [5], or KRAFT [12]. All these works have focused on the 'Multi-Agent Information System' or MAIS.

The second MAS type gathers management assistant agents depending on the actors' needs. In this range, agents are expected to be flexible, pro-active and reactive regarding the user requirements [7],[11].

The new trend of using MAS in knowledge engineering is to associate agents to knowledge structures based on ontologies. This association allows MAS to support the knowledge management process but the issues of knowledge distribution and ontology consistency with MAS have not been solved yet. The next section will present the related work concerning agent approaches using ontologies in knowledge engineering.

2.2. *Ontologies to Support the Knowledge Modeling*

Ontology is an object of Artificial Intelligence that has become a mature powerful conceptual tool of Knowledge Modeling [4]. It provides a coherent base to build on, and a shared reference to align with, in the form of a consensual conceptual vocabulary on which one can build descriptions and communication acts.

Knowledge that is created in engineering projects needs to be defined precisely in order to be useful in an information system. Ontology provides a vocabulary and a semantic that enable the processing of knowledge related to a specific domain. Ontology is a set of items and their specific meanings. It gives definitions and indicates how concepts are connected to each other. These connections form a structure on the defined domain and clarify the possible meanings of the items [20]. Therefore, domain ontology includes the specific concepts of a given domain. It describes the entities, properties and the way they can be related to each other. These ontologies are meant to be re-used in the same domain, in new but similar applications. These ontologies are said to be contextual [16] when the concept properties evolve according to the situation.

2.3. *Interests of the Ontologies in MAS*

The idea of using domain ontologies in an agent system aims at reusing pieces of the domain Knowledge to lead agents to share their information. Indeed in a MAS, several agents interact or work together to carry out common goals [19]. The coordination between agents depends on the process and knowledge they use to achieve their global goals. The domain ontology provides a section of the knowledge world that is essential for the agent to carry out its tasks [6].

Some research works like Buccafurri [3] and Wooldridge [21] use the ontology to give to the agents an internal representation of both interests and behaviour of their associated human users. Other works use ontology to help agents to choose the most promising agents to be contacted for knowledge-sharing goals [6], [4]. Generally, these systems have been designed to prevent the agents from having access to the ontologies of other agents; they ensure an individualistic view of agents' societies. This is the viewpoint of most of the so-called BDI approaches [17][9]. Another interesting approach that has been adopted to design MAS is related to the agent community, where agents automatically build their ontologies by observing the users' actions [2]. Indeed, the agents are able to automatically extract logical rules that represent the user behavior and/or causal implications among events due to the definition of the user interests described with the ontology.

In addition, Guerin [10] and Singh [18] propose to design their MAS in adopting a "social" view of agent communities, where it is assumed that the ontology of each agent is, even partially, accessible for each other agent. In the next section, we will propose the architecture of MAS using a common ontology for all agents. Then, we will present the mechanism of knowledge distribution between agents based on a semantic analysis.

3. Overview of Semknow (Semantic & Knowledge)

In this section we present models that enable to build the knowledge management system dedicated to heterogeneous and distributed information management during engineering projects.

3.1. *Agents and Their Knowledge Worlds*

In a knowledge management system, agents are in a complex information world. To handle and transform information into knowledge, they have to identify it by using models, extract it from different sources, annotate it by respecting a knowledge structure, store it, update it and share it with the users.

We use the OWL-Lite (Web Ontology Language) to build ontologies used in SemKnow (Semantic & Knowledge). OWL-Lite provides a simple triple model based on XML syntax to describe information and their relations. Figure 1 shows an annotation example built by the agents.

```
<owl:Class rdf:ID="Bicycle">
<owl:Class rdf:ID="Mountain Bike">           Subsumption link between Concepts
  <rdfs:subClass rdf:resource ="#Bicycle"/>
</owl:Class>
...
<owl:ObjectProperty rdf:ID="IsComposedBy">     Relation Description
  <rdfs:domain rdf:resource="#Bicycle"/>
  <rdfs:range rdf:resource="#SeatTubeLength"/>
</owl:ObjectProperty>
...                                             Extract of the ontology 'CyclingOnto'
<rdf:description rdf:about="https://acsp.utbm.fr/MB2352.cad">
  <CyclingOnto:Name>SpeedMax CF 9.0<CyclingOnto:Name />     Extract of the annotation of
  <CyclingOnto:SeatTubeLength>550<CyclingOnto:SeatTubeLength />  the ressource 'MB2352.cad'
  <CyclingOnto:TopTubeLength>570<CyclingOnto:TopTubeLength />
  <CyclingOnto:HeadTubeLength>125<CyclingOnto:HeadTubeLength />
  ...
</rdf:description>
```

Figure 1. Example of an annotation builds by the agents

The SemKnow system uses several ontologies built by the professional actors. Each ontology describes a domain (project management, mechanical design, Ergonomics, etc.). The agents are connected to several information systems (e-groupware, project management platform, CAD system etc…) and extract the information to these sources by transforming the ontologies into SQL queries. Thus SemKnow manages the heterogeneous and distributed knowledge with RDF annotations based on several shared ontologies. SemKnow is not a library of indexing documents but a knowledge base with annotations describing information sources with their organization contexts. The SemKnow has to support the knowledge sharing for the users' organization (project team). The problem is to handle the knowledge of the organization and to ensure the distribution of the relevant knowledge to each agent (either human or artificial). In the following section, we present the SemKnow architecture and the mechanisms of knowledge sharing among the agents.

3.2. SemKnow Architecture

Knowledge agents are a part of cognitive and intelligent agents. They constitute a coupled network of agents that worked together to achieve the same objective i.e. to support the knowledge management process by providing full range of functionalities like extracting, annotating, storing, updating and sharing knowledge [15].

The MAS architecture is a structure of an agent network with different types of agents and different relationships between them [8], [9]. The SemKnow architecture starts from the highest level of abstraction with the description of agent societies and goes down to the description of the roles, interactions and responsibilities of the agents.

The proposed approach to design MAS is based on an organizational approach like the A.G.R model used in AALAADIN, OPERA and methodologies like GAIA or TROPOS or RIOCC. Thus the SemKnow architecture is tackled as a human society in terms of role, skill and relationships.

The main objective of the SemKnow system is to manage heterogeneous and distributed knowledge coming from different information sources and used by professional actors. The second objective is to permanently evaluate this knowledge in order to delete obsolete knowledge. The third objective is to assist users in the reuse of knowledge by proposing a decision support. Considering these objectives we have defined four main functionalities for the system:

- To allow users to describe their knowledge domain with a semantic approach (i.e. a characterization of concepts and their relations);
- To extract knowledge from different information sources;
- To update and validate the knowledge base with the users in order to avoid broadcasting wrong information;
- To assist the user in the reuse of knowledge.

Figure 2 gives an overview of the SemKnow architecture.

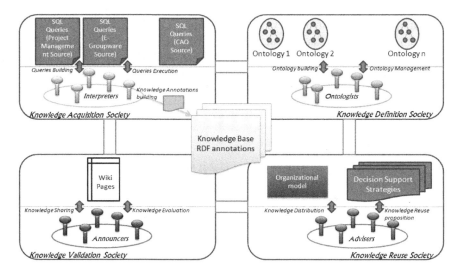

Figure 2. The SemKnow Architecture

In the high level of abstraction we have four agent societies supporting the four functionalities of the SemKnow platform.

The first society of agents, called "ontologists", manages the different ontologies (knowledge models) build by the users. The proposed interface looks like protégé 2000 and the agents help the professional actors to build domain ontologies. The "ontologists" generate ontologies in the OWL-Lite format. There is one ontologist agent for each ontology. The Ontologists also manage the whole ontologies by ensuring the consistency (ontology alignment process and research of similar concepts) between the different ontologies.

The second society of agents is the "Interpreters". Those agents extract knowledge by transforming ontologies in query models using the language SQL. Afterwards, they request the databases of different information sources (E-Groupware, Project Management, CAO, etc.) to extract knowledge. There is one agent for each information source. The interpreters annotate knowledge by giving an organizational context (creator's role, information source, name of the project, etc.). The interpreters build the knowledge base with RDF files.

The third society of agents, called the "Announcers", aims at evaluating, validating and updating knowledge by using a semantic Wiki. The users can consult knowledge by reading Wiki pages and can modify, approve or reject it. The reader can find more details about this process in [14].

The fourth society of agents is the "Advisers". Those agents aim at providing a decision-making support for the users. There is one adviser for each professional actor. The advisers use an organizational model to monitor the actors' actions. This model describes the users' roles, collaborations and activities during a project. Thus, by requesting the knowledge base i.e. the annotations (Knowledge and organizational context), the advisers can alert and propose to users knowledge that has been already stored for a similar past activity, a similar role and a similar information system.

We are going to focus on the work of the Interpreters society i.e. how the agents extract knowledge from the different information sources by using the domain ontologies and transformation rules.

4. Knowledge Extraction by the Interpreters Society

To extract knowledge from the database of business applications the interpreters have to apply an algorithm to extract and translate information of databases according to the OWL ontologies. Hierarchical information is extracted from matches found in the relational dataset. The agents apply some corresponding rules to detect similarity between the structure of the databases tables and the ontologies. The tables 1 and 2 present the correspondence between database components/properties and ontology concepts/relationships.

Table 1: Component/Concept correspondence

Database component	OWL concept
Table	Class
Column	Functional Propoerty
Row	OWL individual
Column Metadata:	OWL property restriction:
Data Type	-AllValues From restriction
Mandatory/Not nullable	-Cardinality () Restriction
Nullable	-maxCardinality() Restriction

Table 2: Property/Relationship correspondence

Database property	OWL relationship
NOT NULL	owl:minCardinality rdf:datatype=''& xsd:Int''1/
UNIQUE	owl:InversFunctionalProperty
CHECK	owl:hasValue
FOREIGN KEY	owl:objectProperty

We have experimented transformation rules with four different information sources (an e-Groupware, a project management system, a risk analysis system and a CAO platform). The Interpreters apply 10 transformation rules to go from the ontology to the SQL model. These rules are described below:

_R1_ The name of the ontology is the name of a project

_R2_ A class is a condition on the name of product or element of product

_R3_ If a class has one or more DatatypeProperty so we the condition is ParametersName in ('DatatypeProperty1', 'DatatypeProperty2', 'DatatypeProperty3'...)

_R4_ Forming conditions: We look at the range of each DatatypeProperty

- If it is an attribute positiveInteger then the condition is check> 0

- If it is a DataRange which includes a list so check attribute in [value1,value2 ...]

- If there are restrictions on DatatypeProperty and it is not a cardinality restriction as HasValue so the condition is: check attribute = value

- Inverse functional property is the "Disctinct" constraint

- Required property mean check fields not null (not 0 or not = "")

_R5_ Store conditions in the list X.

R6 We have to mention that we have a table "Matching table" which stores all tables, columns of all databases and the equivalent variable. For instance, variable '$ParametersName' corresponds to column 'nom_variable' in the table 'variables' in the database 'ACSP', the same variable '$ParametersName' corresponds also to column 'nom_parametre' in the table 'parametres in the database 'KrossRoad'.

As we have in a table above the list of tables, we can search the tables of the database on which we applied our ontology and store them in the list Y

R7 Search the primary and foreign keys of the tables in the list (the relations between them) and store these relations in a table.

R8 We search all times the values of the following variables: $ParametersName, $ParametersValue, $ProjectName, $ProjectDesciption, $ProjectStartDay, $ProjectEndDay, $PersonName, $PersonRole. So, we search in the "Matching table" the equivalence of these variables in the database we work on and store them in the list Z.

R9 Build the query:

- **Select** the columns in the list Z proceeded by the first letter of the domain table. E.g.: p.nom, And followed by a comma

- **From** all tables stored in the list Y followed by the first letter as naming the table by a letter

- **Where** all the relationship stored in a table to make the connections between the tables + all conditions sored in the list X separated by "and".

R10 we repeat these rules for each class int the ontology who have datatypeproperty and we do "Union" between these queries to get one result at the end

We illustrate the mechanism used by the interpreters for the knowledge extraction with an ontology concerning the cycling domain.

The cycling ontology describes the cycling world. The ontology defines a vocabulary and a semantic to structure, organize, detail all the characteristics of a bike, all the roles of the professional actors during the development project of a new bike and all the processes used to develop and industrialize a bicycle.

By considering the bike ontology (figure 3), business expert can add in his ontology general concepts. He can add also instances of concepts in order to filter the results of the ontology. So, instances in this ontology play the role of conditions. For instance, in the bike ontology business expert wants to know the right values of parameters to design a new bike. But he does not want any bike, so he specifies his search by indicating the instance "Ultimate CF SLX 9.0 SL" as a name of the bike. In addition, he specifies the instance "Canyon One Four SLX" which means that the bike must have this name as fork, also a "Mavic R-Sys SL" as rims and "Selle Italia SLR" as saddle. Thus these conditions can be considered as filter of the result as expected. On the other side, he can create general concepts in his ontology as the concept "brakes", "chainrings", "frame", etc. without specifying conditions on these concepts.

In the proposed ontology there are relations between existing concepts. For example, the relation "is_stopped_by" links the instance "Ultimate CF SLX 9.0 SL" to the concept "brakes", the relation "is_composed_by" links the instance "Ultimate CF SLX 9.0 SL" to the concept "frame", the relation "is_fitted_into" links the concept "wheel" to the concept "frame", etc.

➢ Ultimate CF SLX 9.0 SL (Bike)
　　➢ Brakes
　　➢ Chainrings
　　➢ Canyon One Four SLX (Fork)
　　➢ Frame
　　　　➢ Seat tube length
　　　　➢ Top tube length
　　　　➢ Head tube length
　　　　➢ Head tube angle
　　　　➢ Seat tube angle
　　　　➢ Chainstay length
　　　　➢ Wheel base
　　　　➢ Stack
　　　　➢ Reach
　　➢ Handlebar
　　　　➢ Weight
　　　　➢ Width
　　　　➢ type
　　　　➢ Material
　　➢ Headset
　　➢ Wheel
　　　　➢ Hub
　　　　　　➢ Axle
　　　　　　➢ Bearings
　　　　　　➢ Hub shell
　　　　　　　　➢ Hub brakes
　　　　　　　　➢ Gears

➢   Mavic R-Sys SL (Rims)
    ➢   Size
    ➢   Weight
➢   Spokes
    ➢   Material
    ➢   Number of spokes
➢   Nipples
➢   Selle Italia SLR (Saddle)
➢   Rear Derailleur
➢   Front Derailleur
➢   Pedals

Figure 3. An extract of the bike ontology

This ontology is transformed into an SQL query which will return information from business applications. For example if business experts want to know how to design an "Ultimate CF SLX 9.0 SL" bike so the Interpreters agents transform the cycling ontology (figure 3) in specifying these details. The cycling ontology will be transformed into an SQL query such as the example below. The information returned with this query becomes knowledge and stored in RDF files as shown in Figure 1 in the annotation part.

```
select v.nom_variable, v.valeur_variable, pc.nom, pc.description, pc.datedebut,
pc.datefin, pp.nom, pp.role
from variables v, produit_element pe, produit p, projet pc, personne pp
where v.id_produit = pe.n_produit
and v.id_element = pe.n_element
and pe.n_produit = p.id_produit
and p.id_projet = pc.id_projet
and pc.id_personne = pp.id_personne
and v.nom_variable in ('size','weight')
and pe.nom_pe like'% Mavic R-Sys SL%'
and p.nom_produit like '%ultimate%'
and pc.id_projet like '%bike%'
UNION
select ...
from ...
where ...
and v.nom_variable in ('Weight', 'Width', 'type', 'material')
and pe.nom_pe like'% Handlebar%'
and p.nom_produit like '%ultimate%'
and pc.id_projet like '%bike%'
UNION ...
```

Figure 4. Extract of the generated SQL query

By using those rules, we have succeeded in extracting 58% of the concepts defined in the ontologies. It is more than previous work like in [2] but we have to more improve this result. Indeed it depends on the structure of databases. If the database has not relevant relations we obtain a multitude of results which are not relevant.

The loss of semantics is due to the fact that some of the concepts and relations defined in the ontology have no equivalence in the database (table or relations). In the SemKnow platform, the professional actors build their own ontology from their professional expertise. We observed that when actors had a good acquaintance with the professional software tool he/she used, then he/she properly defined ontology with concepts close from the database structure and the agents obtained good knowledge extraction results.

5. Conclusion

This paper has presented the architecture of the SemKnow platform with four agent societies which cover the knowledge management process. It has focused on the knowledge extraction that is carried out by the Interpreters agents. The next step of this research will be to improve the obtained results according to two ways; the first will be to find new transformation rules to decrease the loss of semantics and the second concerns the development of new models used by the ontologists agent society, to better assist the human experts in the building of domain ontologies that with similar concepts that they use when they work with their professional software tools.

References

1. A. Abecker, A. Bernardi, and L. van Elst, Agent technology for distributed organizational memories. *In Proceedings of the 5th International Conference On Enterprise Information Systems*, Vol. 2, pages 3–10 (2003)
2. I. Astrova, N. Korda, and A. Kalja, Storing OWL Ontologies in SQL Relational Databases. World Academy of Science, *Engineering and Technology* 29 (2007)
3. F. Buccafurri, D. Rosaci, G.M.L. Sarne´ , D. Ursino, An agent-based hierarchical clustering approach for e-commerce environments, in: *Proceedings of the Third E-Commerce and Web Technologies* (EC-Web 2002), Aix-en-Provence, France, *Lecture Notes in Computer Science, Springer*, Berlin, pp. 115–118 (2002)
4. G. Beydoun, N. Tran, G. Low and B. Henderson-Sellers, Foundations of Ontology-Based MAS Methodologies, *Agent-Oriented Information Systems III, SpringerLink,* Volume 3529 (2006)

5. M. Bonifacio, P. Bouquet, and P. Traverso. Enabling distributed knowledge management. *Managerial and technological implications. Novatica and Informatik/Informatique*, 3(1):23. 29 (2002)

6. M. C. Bravo, J. Perez, V. J. Sosa, A. Montes, G. Reyes, Ontology Support for Communicating Agents in Negotiation Process, *International Conference on Hybrid Intelligent System*, 12p. (2005)

7. P.A Champin, Y. Prie, A. Mille, MUSETTE: Modelling USEs and Tasks for Tracing Experience, *Proc. From structured cases to unstructured problem solving episodes - WS 5 of ICCBR'03*, Trondheim (NO) , NTNU, Trondheim (NO) , pp279-286 (2003)

8. F. Gandon, L. Berthelot , Dieng-Kuntz R., A Multi-Agent Platform for a Corporate Semantic Web, AAMAS 2002, *6th International Conference on Autonomous Agents*, 5th International Conference on Multi-Agents Systems, 9th International Workshop on Agent Theories Architectures and Languages Eds Castelfranchi C., Johnson W.L., p. 1025-1032, July 15-19, Bologna, Italy (2002).

9. N. Griffiths, M. Luck, Cooperative plan selection through trust, in: *Proceedings of the Ninth European Workshop on Modelling Autonomous Agents in a Multi-Agent World: Multi-Agent System Engineering* (MAAMAW-99), vol. 1647, Springer, Berlin, pp. 162.174 (1999).

10. F. Guerin, J. Pitt, Denotational semantics for agent communication language, in: AGENTS .01: *Proceedings of the Fifth International Conference on Autonomous Agents, ACM Press*, pp. 497.504 (2001)

11. M. Klusch, Intelligent Information Agents: Agent-based Information Discovery and Management in the Internet, *Springer* (1999).

12. J. Kyengwhan, Jung-Jin Yang, Knowledge Description Model for MAS Utilizing Distributed Ontology Repositories, in *Agent Computing and Multi-Agent Systems, SpringerLink*, Volume 4088 (2006)

13. D. Monticolo, I. Lahoud, E. Bonjour, F. Demoly, SemKnow: A Multi-Agent Platform to Manage Distributed Knowledge by using Ontologies, *Lecture Notes in Engineering and Computer Science: Proceedings of The International MultiConference of Engineers and Computer Scientists 2012*, IMECS 2012, Vol.1, 14-16 March, Hong Kong, pp. 58-62 (2012)

14. D. Monticolo, S. Gomes, Collaborative Knowledge Evaluation with a Semantic Wiki: WikiDesign, in *the International Journal of e-Collaboration*, Volume 7, Issue 3. 12 pages (2011)

15. M. R. Lee, T. T. Chen, Revealing Research Themes and Trends in KnowledgeManagement: from 1995 to 2010, in *Knowledge-Based Systems*, (2011)

16. P. Pouquet, C-owl: Contextualizing ontologies. *In Proceedings of the 2nd International Semantic Web Conference (ISWC2003)*, 20-23 October 2003, Sundial Resort, Sanibel Island, Florida, USA., (2003)

17. A. Rao, M. Georgeff, Decision procedures of bdi logics, *J. Logic Comput.* 8 (3) (1998).

18. B. Singh. Interconnected Roles (IR): A Coordination Model. *Technical Report CT-084-92*, MCC, (1992).
19. K. Sycara, In-Context Information Management through Adaptive Collaboration of Intelligent Agents, p53-77 *in the book "Intelligent Information Agent: Agent-Based Information Discovery and Management on the Internet", Matthias Klusch*, Springer, (1999)
20. M. Uschold, "Building ontologies: towards an Unified Methodology. *In proceedings of the 16th conference of the British Computer Society Specialist Group and Expert Systems*, Cambridge UK, (1996).
21. M. Wooldridge, Reasoning About Rational Agents, *MIT Press*, Cambridge, MA, (2000)

AN INTELLIGENT TRAIN MARSHALING BASED ON THE PROCESSING TIME CONSIDERING GROUP LAYOUT OF FREIGHT CARS

YOICHI HIRASHIMA

*Faculty of Information Science and Technology, Osaka Institute of Technology,
1-79-1, Kita-yama, Hirakata City, Osaka, 573-0196, Japan
E-mail: hirash-y@is.oit.ac.jp
www.is.oit.ac.jp*

In this chapter, an intelligent method for generating marshaling plan of freight cars in a train is introduced. Initially, freight cars are located in a freight yard by the random layout, and they are moved and lined into a main track in a certain desired order in order to assemble an out bound train. Based on the processing time, Marshaling plans are obtained by a reinforcement learning system. In order to evaluate the processing time, the total transfer distance of a locomotive and the total movement counts of freight cars are simultaneously considered. Moreover, by grouping freight cars that have the same destination, candidates of the desired arrangement of the outbound train is extended. This feature is considered in the learning algorithm, so that the total processing time is reduced. Then, the order of movements of freight cars, the position for each removed car, the layout of groups in a train, the arrangement of cars in a group and the number of cars to be moved are simultaneously optimized to achieve minimization of the total processing time for obtaining the desired arrangement of freight cars for an outbound train.

Keywords: Scheduling, Container Transfer Problem, Q-Learning, Freight train, Marshaling

1. Introduction

Logistics with freight train has a growing importance in recent years, because railway logistics is known to have smaller environmental load as compared to goods transportation with trucks.[1] Train marshaling operation at freight yard is required to joint several rail transports, or different modes of transportation including rail. Transporting goods are carried in containers, each of which is loaded on a freight car. A freight train is consists of several freight cars, and each car has its own destination. Thus, the train driven by a locomotive travels several destinations delivering corresponding freight cars at each freight station. In addition, since freight trains can transport goods only between railway stations, modal shifts are required for area that has no railway. In intermodal transports including rail, containers carried into the station are loaded on freight cars and located at the freight yard in the arriving order. The initial layout of freight cars is thus random. For efficient shift in assembling outbound train, freight cars must be rearranged before coupling to the freight train. In general, the rearrangement process is conducted in a freight yard that consists of a main-track and several sub-tracks. Freight cars are initially placed on sub tracks, rearranged, and lined into the main track. This series of operation is called marshaling, and several methods to solve the marshaling problem have been proposed.[2,3] Also, many similar problems are treated by mathematical programming and genetic algorithm,[4-7] and some analyses are conducted for computational complexities.[7,8] However, these methods do not consider the processing time for each transfer movement of freight car that is moved by a locomotive. In our research group, methods that generates each transfer movement of freight car in a marshaling are designed based on the movement counts of freight cars, and based on the transfer distance of a locomotive.[9] By extending these methods, a rainforcement learning method that can obtain and improve marshaling plans based on the processing time has been proposed.[10]

In this chapter the method for generating marshaling plan of freight cars in a train is introduced. In the method, marshaling plans based on the processing time can be obtained by a reinforcement learning system. A movement of a freight car consists of 4 elements: 1. moving a locomotive to the car to be transferred, 2. coupling cars with the locomotive, 3. transferring cars to their new position by the locomotive, and 4. decoupling the cars from the locomotive. The processing times for elements 1, 3 are determined by the transfer distance of the locomotive, the weight of the train, and the performance of the locomotive. The total processing time for elements 2, 4 is determined by the number of movements of freight cars. Thus, the transfer distance of the locomotive and the number of movements of freight cars are simultaneously considered, and used to evaluate and minimize the processing time of marshaling for obtaining the desired layout of freight cars for an outbound train. The total processing time of marshaling is considered by using a weighted cost of a transfer distance of the locomotive and the number of movements of freight cars. Then, the order of movements of freight cars, the position for each removed car, the arrangement of cars in a train and the number of cars to be moved are simultaneously optimized to achieve minimization of the total processing time. The *original* desired arrangement of freight cars in the main track is derived based on the destination of freight cars. In the proposed method, by grouping freight cars that have the same destination, several desirable positions for each freight car in a group are generated from the original one, and the optimal group-layout that can achieve the smallest processing time of marshaling is obtained by autonomous learning. Simultaneously, the optimal sequence of car-movements as well as the number of freight cars that can achieve the desired layout is obtained by autonomous learning. Also, the feature is considered in the learning algorithm, so that, at each arrangement on sub track, an evaluation value represents the smallest processing time of marshaling to achieve the best layout on the main track. The learning algorithm is derived based on the Q-Learning,[11] which is known as one of the well established realization algorithm of the reinforcement learning. In order to show effectiveness of the proposed method, computer simulations are conducted for several methods.

2. Problem Description

The yard consist of 1 main track and m sub tracks. Define k as the number of freight cars placed on the sub tracks, and they are carried to the main track by the desirable order based on their destination. In the yard, a locomotive moves freight cars from sub track to sub track or from sub track to main track. The movement of freight cars from sub track to sub track is called removal, and the car-movement from sub track to main track is called rearrangement. For simplicity, the maximum number of freight cars that each sub track can have is assumed to be n, the ith car is recognized by an unique symbol c_i ($i = 1, \cdots, k$). Fig.1 shows the outline of freight yard in the case $k = 30, m = n = 6$. In the figure, track T_m denotes

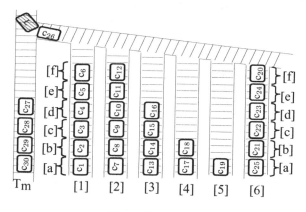

Fig. 1.   Freight yard

the main track, and other tracks [1], [2], [3], [4], [5], [6] are sub tracks. The main track is linked with sub tracks by a joint track, which is used for moving cars between sub tracks, or for moving them from a sub track to the main track. In the figure, freight cars are moved from sub tracks, and lined in the main track by the descending order, that is, rearrangement starts with c_{30} and finishes with c_1. When the locomotive L moves a certain car, other cars locating between the locomotive and the car to be moved must be removed to other sub tracks. This operation is called removal. Then, if $k \leq n \cdot m - (n - 1)$ is satisfied for keeping adequate space to conduct removal process, every car can be rearranged to the main track.

In each sub track, positions of cars are defined by n rows. Every position has unique position number represented by $m \cdot n$ integers, and the position number for cars at main track is 0. Fig.2 shows an example of position index for $k = 30, m = n = 6$ and the layout of cars for fig.1D

[f]	31	32	33	34	35	36
[e]	25	26	27	28	29	30
[d]	19	20	21	22	23	24
[c]	13	14	15	16	17	18
[b]	7	8	9	10	11	12
[a]	1	2	3	4	5	6
	[1]	[2]	[3]	[4]	[5]	[6]

Position index

[f]	c_6	c_{12}				c_{20}
[e]	c_5	c_{11}				c_{24}
[d]	c_4	c_{10}	c_{16}			c_{23}
[c]	c_3	c_9	c_{15}			c_{22}
[b]	c_2	c_8	c_{14}	c_{18}		c_{21}
[a]	c_1	c_7	c_{13}	c_{17}	c_{19}	c_{25}
	[1]	[2]	[3]	[4]	[5]	[6]

Yard layout

Fig. 2.   Example of position index and yard state

In Fig.2, the position "[a][1]" that is located at row "[a]" in the sub track "[1]" has the position number 1, and the position "[f][6]" has the position number 36. For unified representation of layout of car in sub tracks, cars are placed from the row "[a]" in every track, and newly placed car is jointed with the adjacent freight car. In the figure, in order to rearrange c_{25}, cars $c_{24}, c_{23}, c_{22}, c_{21}$ and c_{20} have to be removed to other sub tracks. Then, since $k \leq n \cdot m - (n - 1)$ is satisfied, c_{25} can be moved even when all the other cars are placed in sub tracks.

In the freight yard, define $x_i (1 \leq x_i \leq n \cdot m, i = 1, \cdots, k)$ as the position number of the car c_i, and $s = [x_1, \cdots, x_k]$ as the state vector of the sub tracks. For example, in Fig.2, the state is represented by $s = [1, 7, 13, 19, 25, 31, 2, 8, 14, 20, 26, 32, 3, 9, 15, 21, 4, 10, 5, 36, 12, 18, 24, 30, 6, 0, 0, 0, 0, 0]$. A trial of the rearrange process starts with the initial layout, rearranging freight cars according to the desirable layout in the main track, and finishes when all the cars are rearranged to the main track.

3. Desired Layout in the Main Track

In the main track, freight cars that have the same destination are placed at the neighboring positions. In this case, removal operations of these cars are not required at the destination regardless of layouts of these cars. In order to consider this feature in the desired layout in the main track, a group is organized by cars that have the same destination, and these cars can be placed at any positions in the group. Then, for each destination, make a corresponding group, and the order of groups lined in the main track is predetermined by destinations. This feature yields several desirable layouts in the main track.

Fig.3 depicts examples of desirable layouts of cars and the desired layout of groups in the main track. In the figure, freight cars $c_1, \cdots, c_6$ to the $destination_1$ make $group_1$, $c_7, \cdots, c_{18}$ to the $destination_2$ make $group_2$, $c_{19}, \cdots, c_{25}$ to the $destination_3$ make $group_3$, and $c_{26}, \cdots, c_{30}$ to the $destination_4$ make $group_4$. $Groups_{1,2,3,4}$ are lined by ascending order in the main track, which make a desirable layout. In the figure, examples of layout in $group_1$ are in the dashed square.

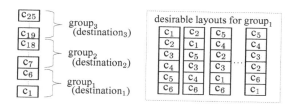

Fig. 3.   Example of groups

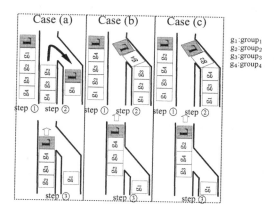

Fig. 4.   Group layouts

Also, the layout of groups lined by the reverse order do not yield additional removal actions at the destination of each group. Thus, in the proposed method, the layout lined groups by the reverse order and the layout lined by decoupling order from both ends of the train are regarded as desired layouts. Fig.4 depicts examples of material handling operation for extended layout of groups at the destination of $group_1$. In the figure, step ① shows the layout of the incoming train. In case (a), cars in $group_1$ are separated at the main track, and moved to a sub-track by the locomotive L at step ②. In cases (b),(c), cars in $group_1$ are carried in a sub-track, and $group_1$ is separated at the sub-track. In the cases, $group_1$ can be located without any removal actions for cars in each group. Thus, these layouts of groups are regarded as candidate for desired one in the learning process of the proposed method.

4. Direct Rearrangement

When rearranging car that has no car to be removed on it is exist, its rearrangement precede any removals. In the case that several cars can be rearranged without a removal, rearrangements are repeated until all the candidates for rearrangement requires at least one removal. If several candidates for rearrangement require no removal, the order of selection is random, because any orders satisfy the desirable layout of groups in the main track. In this case, the arrangement of cars in sub tracks obtained after rearrangements is unique, so that the movement counts of cars has no correlation with rearrangement orders of cars that require no removal. This operation is called direct rearrangement. When a car in a certain sub track can be rearrange directly to the main track and when several cars located adjacent positions in the same sub track satisfy the layout of group in main track, they are jointed and applied direct rearrangement.

Fig.5 shows an example of arrangement in sub tracks existing candidates for rearranging cars that require no removal. At the top of figure, from the left side, a desired layout of cars and groups, the initial layout of cars in sub tracks, and the position index in sub tracks are depicted for $m = n = 4, k = 9$. c_1, c_2, c_3, c_4 are in $group_1$, c_5, c_6, c_7, c_8 are in $group_2$, and $group_1$ must be rearranged first to the main track. In each group, any layouts of cars can be acceptable. In both cases, c_2 in step 1 and c_3 in step 3 are applied the direct rearrangement. Also, in step 4, 3 cars c_1, c_4, c_5 located adjacent positions are jointed and moved to the main track by a direct rearrangement operation. In addition, at step 5 in case 2, cars in $group_2$ and $group_3$ are moved by a direct rearrangement, since the positions of c_7, c_8, c_6, c_9 are satisfied the desired layout of groups in the main track.

In case 1 of the example, the rearrangement order of cars that require no removal is c_1, c_2, c_3, c_4, and in case 2, the order is c_3, c_2, c_1, c_4. Although 2 cases have different orders of rearrangement, the arrangements of cars in sub tracks and the numbers of movements of cars have no difference.

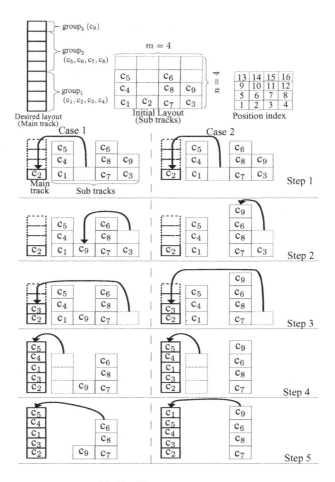

Fig. 5.   Direct rearrangements

5. Rearrangement Process

The rearrangement process for cars consists of following 6 operations:

(**I**)   selection of a layout of groups in the main track, and rearrangement for all the cars that can apply the direct rearrangement into the main track,

(**II**)   selection of a freight car to be rearranged into the main track,

(**III**) selection of a removal destinations of the cars in front of the car selected in (I),

(**IV**) selection of the number of cars to be moved,

(V) removal of the cars to the selected sub-track,

(VI) rearrangement of the selected car.

These operations are repeated until one of desirable layouts is achieved in the main track, and a series of operations from the initial state to the desirable layout is define as a trial.

Now, define h as the number of candidates of the desired layout of groups. Each candidate in operation (I) is represented by u_{j_1} $(1 \leq j_1 \leq h)$.

In the operation (II), each group has the predetermined position in the main track. The car to be rearranged is defined as c_T, and candidates of c_T can be determined by excluding freight cars that have already rearranged to the main track. These candidates must belong to the same group.

Also, define r as the number of groups, g_l as the number of freight cars in $\text{group}_l (1 \leq l \leq r)$, and $u_{j_2}(h + 1 \leq j_2 \leq h + g_l)$ as candidates of c_T.

In the operation (III), the removal destination of car located on the car to be rearranged is defined as c_M. Then, defining $u_{j_3}(h + g_l + 1 \leq j_3 \leq h + g_l + m - 1)$ as candidates of c_M, excluding the sub-track that has the car to be removed, and the number of candidates is $m - 1$.

In the operation (IV), defining p as the number of removal cars required to rearrange c_T, and defining q as the number of removal cars that can be located the sub-track selected in the operation (III), the candidate numbers of cars to be moved are determined by $u_{j_4}, 2m \leq j_4 \leq 2m + \min\{p, q\} - 1$.

In both cases of Fig.5, the direct rearrangement is conducted for c_2 at step 1, and the selection of c_T conducted at step 2, candidates are $u_{h+1} = [1], u_{h+2} = [4]$, that is, sub-tracks where cars in group_1 are located at the top. u_{h+3}, u_{h+4} are excluded from candidates. Then, $u_{h+2} = [4]$ is selected as c_T. Candidates for the location of c_T are $u_{h+5} = [1], u_{h+6} = [2], u_{h+7} = [3]$, sub-tracks [1],[2], and [3]. In case 1, $u_6 = [2]$ is selected as c_M, and in case 2, $u_{h+7} = [3]$ is selected. After direct rearrangements of c_3 at step 3 and c_1, c_4, c_5 at step 4, the marshaling process is finished at step 5 in case 2, whereas case 1 requires one more step in order to finish the process. Therefore, the layout of cars and groups in the main track, the number of cars to be moved, the location the car to be rearranged and the order of rearrangement affect the total movement counts of freight cars.

6. Processing Time for a Movement of Locomotive

6.1. *Transfer distance of locomotive*

When a locomotive transfer freight cars, the process of the unit transition is as follows: (E1). starts without freight cars, and reaches to the joint track, (E2) restart in reverse direction to the target car to be moved, (E3). couples them, (E4) pull out them to the joint track, (E5) restart in reverse direction, and transfers them to the indicated location, and (E6) decouples them from the locomotive. Then, the transfer distance of locomotive in (E1), (E2), (E4) and (E5) is defined as D_1, D_2, D_3 and D_4, respectively. Also, define the unit distance of a movement for cars in each sub track as $D_{\min_v}$, the length of joint track between adjacent sub tracks, or, sub track and main track as $D_{\min_h}$. The location of the locomotive at the end of above process is the start location of the next movement process of the selected car. Also, the initial position of the locomotive is located on the joint track nearest to the main track.

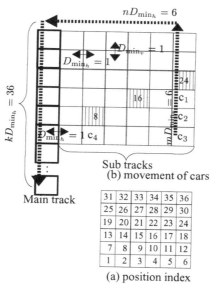

Fig. 6.   Calculation of transfer distance

Fig.6 shows an example of transfer distance. In the figure, $m = n = 6$, $D_{\min_v} = D_{\min_h} = 1$, $k = 18$, (a) is position index, and (b) depicts movements of locomotive and freight car. Also, the locomotive starts from position 8, the target is located on the position 18, the destination of the target is 4, and the number of cars to be moved is 2. Since the locomotive moves without freight cars from 8 to 24, the transfer distance is $D_1 + D_2 = 12$ ($D_1 = 5$, $D_2 = 7$), whereas it moves from 24 to 16 with 2 freight cars, and the transfer distance is $D_3 + D_4 = 13$ ($D_3 = 7$, $D_4 = 6$).

6.2. *Processing time for the unit transition*

In the process of the unit transition, the each time for (E3) and (E6) is assumed to be the constant t_E.

The processing times for elements (E1), (E2), (E4) and (E5) are determined by the transfer distance of the locomotive $D_i (i = 1, 2, 3, 4)$, the weight of the freight cars W moved in the process, and the performance of the locomotive. Then, the time each for (E1), (E2), (E4) and (E5) is assumed to be obtained by the function $f(D_i, W)$ derived considering dynamics of the locomotive, limitation of the velocity, and control rules. Thus, the processing time for the unit transition t_U is calculated by $t_U = t_E + \sum_{i=1}^{2} f(D_i, 0) + \sum_{j=4}^{5} f(D_j, W)$. The maximum value of t_U is define as $t_{\max}$ and is calculated by

$$
\begin{aligned}
t_{\max} = {} & t_E + f(kD_{\min_v}, 0) + f(mD_{\min_h}, 0) \\
& + f(mD_{\min_h} + n, W) + f(kD_{\min_v}, W)
\end{aligned} \tag{1}
$$

7. Learning Algorithm

Define h as the number of candidates of the desired layout of groups. Each candidate is represented by u_{j_1} $(1 \leq u_{j_1} \leq h)$, and evaluated by $Q_1(u_{j_1})$. Then, $Q_1(u_{j_1})$ is updated by the following equation when one of desired layout is achieved in the main track:

$$
Q_1(u_{j_1}) \leftarrow \max\{Q_1(u_{j_1}), (1 - \alpha)Q_1(u_{j_1}) + \alpha\gamma^l R\}, \tag{2}
$$

where l denotes the total movement counts required to achieve the desired layout, α is learning rate, γ is discount factor, R is reward that is given only when one of desired layout is achieved in the main track.

Define $s(t)$ as the state at time t, r_M as the sub track selected as the destination for the removed car, p_M as the number of removed cars, q as the movement counts of freight cars by direct rearrangement, and s' as the state that follows s. Also, Q_2, Q_3, Q_4 are defined as evaluation values for (s_1, u_{j_2}), (s_2, u_{j_3}), (s_3, u_{j_4}), respectively, where $s_1 = s, s_2 = [s, c_T], s_3 = [s, c_T, r_M]$. $Q_2(s_1, u_{j_2})$, $Q_3(s_2, u_{j_3})$ and $Q_4(s_3, u_{j_4})$ are updated by following rules:

$$Q_2\,(s_1, c_T) \leftarrow \max_{u_{j_3}} Q_3(s_2, u_{j_3}), \tag{3}$$

$$Q_3\,(s_2, r_M) \leftarrow \max_{u_{j_4}} Q_4(s_3, u_{j_4}), \tag{4}$$

$$Q_4\,(s_3, p_M) \leftarrow \tag{5}$$

$$\begin{cases} (1-\alpha)Q_4(s_3, p_M) + \alpha[R + \gamma^{q+1}V_1] \\ \text{(next action is rearrangement)} \\ (1-\alpha)Q_4(s_3, p_M) + \alpha[R + \gamma V_2] \\ \text{(next action is removal)} \end{cases}$$

$$V_1 = \max_{u_{j_1}} Q_2(s_1', u_{j_2}),$$
$$V_2 = \max_{u_{j_2}} Q_3(s_2', u_{j_3})$$

where α is the learning rate, R is the reward that is given when one of desirable layout is achieved, and γ is the discount factor that is used to reflect the processing time of the marshaling and calculated by the following equation.

$$\gamma = \delta \frac{t_{\max} - \beta t_U}{t_{\max}}, \ 0 < \beta < 1, 0 < \delta < 1 \tag{6}$$

Propagating Q-values by using eqs.(3)-(6), Q-values are discounted according to the processing time of marshaling. In other words, by selecting the removal destination that has the largest Q-value, the processing time of the marshaling can be reduced.

In the learning stages, each $u_j\,(1 \leq j \leq h + 2m + \min\{p_s, p_d\} - 1)$ is selected by the soft-max action selection method.[12] Probability P for selection of each candidate is calculated by

$$\tilde{Q}_i(s, u_{j_i}) = \frac{Q_i(s, u_{j_i}) - \min\limits_{u} Q_i(s, u_{j_i})}{\max\limits_{u} Q_i(s, u_{j_i}) - \min\limits_{u} Q_i(s, u_{j_i})} \tag{7}$$

$$P(s_i, u_{j_i}) = \frac{\exp(\tilde{Q}_i(s_{j_i}, u_{j_i})/\xi)}{\sum\limits_{u \in u_{j_i}} \exp(\tilde{Q}_i(s_i, u)/\xi)}, \tag{8}$$

$$(i = 1, 2, 3, 4).$$

In the addressed problem, Q_1, Q_2, Q_3, Q_4 become smaller when the number of discounts becomes larger. Then, for complex problems, the difference between probabilities in candidate selection remain small at the initial state and large at final state before achieving desired layout, even after repetitive learning. In this case, obtained evaluation does not contribute to selections in initial stage of marshaling process, and search movements to reduce the transfer distance of locomotive is spoiled in final stage. To conquer this drawback, Q_1, Q_2, Q_3, Q_4 are normalized by eq.(7), and the thermo constant ξ is switched from ξ_1 to ξ_2 ($\xi_1 > \xi_2$) when the following condition is satisfied:

$$[\text{The count of } Q_i(s_{j_i}, u_{j_i})] > \eta,$$
$$\text{s.t. } Q_i(s_{j_i}, u_{j_i}) > 0, \tag{9}$$
$$0 < \eta \le [\text{the number of candidates for } u_{j_i}]$$

where η is the threshold to judge the progress of learning.

8. Computer Simulations

Computer simulations are conducted for $m = 12, n = 6, k = 36$ and learning performances of following 4 methods are compared:

(A) proposed method that evaluates the processing time of the marshaling operation, considering the layout of groups,

(B) a method that evaluates the transfer distance of the locomotive,[9] considering the layout of groups,

(C) a method that evaluates the number of movements of freight cars, considering the layout of groups,

(D) proposed method that evaluates the processing time of the marshaling operation, with single layout of groups,

(E) a method that evaluates the number of movements of freight cars, with single layout of groups.[9]

The initial arrangement of cars in sub tracks is described in Fig.7. In this case, the rearrangement order of groups is group_1, group_2, group_3, group_4. Cars $c_1, \cdots, c_9$ are in group_1, $c_{10}, \cdots, c_{18}$ are in group_2, $c_{19}, \cdots, c_{27}$ are in group_3, and $c_{28}, \cdots, c_{36}$ are in group_4. Other parameters are set as $\alpha = 0.9, \beta = 0.2, \delta = 0.9, R = 1.0, \eta = 0.95, \xi_1 = 0.1, \xi_2 = 0.05$. In methods (C),(E), the discount factor γ is assumed to be constant, and set as $\gamma = 0.9$ instead of the calculation with eq.(6). In method (D), the desired layout of groups in the main track is ascending order from the head, that is, group_1, group_2, group_3, group_4.

The locomotive assumed to accelerate and decelerate the train with the constant force 100×10^3N, and to be 100×10^3kg in weight. Also, all the freight cars have the same weight, 10×10^3kg. The locomotive and freight cars assumed to have the same length, and $D_{\min_v} = D_{\min_h} = 20$m. The velocity of the locomotive is limited to no more than 10m/s. Then, the locomotive accelerates the train until the velocity arrives 10m/s, keeps the velocity, and decelerates until the train stops within the indicated distance. When the velocity does not arrive 10m/s at the half way point, the locomotive starts to decelerate immediately. Then, we set $t_{max} = 462$.

Fig.8 show the results. In Fig.8, horizontal axis expresses the number of trials and the vertical axis expresses the minimum processing time to achieve a desirable layout found in the past trials. Each result is averaged over 20 independent simulations. In Fig.8, the learning performance of methods (A),(C) is better than that of methods (D),(E), because solutions derived by methods (A),(C) use the extended layouts of groups for reducing the total processing time. In method (C), the learning algorithm evaluates the number of movements of freight cars, and is not effective to reduce the total processing time as compared to method (A). In method (B), only the total transfer distance of locomotive is evaluated, so that the total processing time is not improved adequately even if many trials are repeated. Total transfer distances of the locomotive at 1.5×10^6th trial are described in table.1 for each method. Fig.9 shows final arrangements of freight cars generated by the best solutions derived by methods (A) and (D). Since the layout of group is extended, method (A) learns the layout of groups in order to reduce the total processing time, whereas the layout is fixed to the ascending order in method (D).

Table 1. Total processing time

methods	processing time (sec.)		
	best	average	worst
method (A)	5328.06	5376.33	5407.88
method (B)	5331.81	5390.69	5423.99
method (C)	5337.18	5366.46	5416.54
method (D)	5688.26	5763.90	5839.88
method (E)	5788.20	5863.09	5939.08

				c_{19}	c_{32}	c_{31}					
c_{10}		c_{34}	c_{28}	c_{16}	c_4	c_5		c_{25}	c_{29}		c_{36}
c_2	c_{13}	c_8	c_{11}	c_{14}	c_{17}	c_{20}	c_{23}	c_{26}	c_1	c_{33}	c_{35}
c_3	c_6	c_9	c_{12}	c_{15}	c_{18}	c_{21}	c_{24}	c_{27}	c_{30}	c_{22}	c_7

Fig. 7. Initial layout

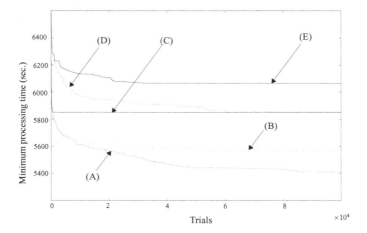

Fig. 8.   Comparison of learning performances

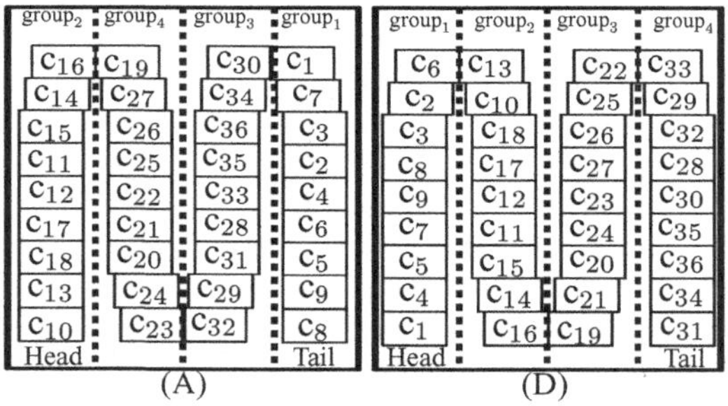

Fig. 9.   Final layouts

9. Conclusions

A new marshaling method for assembling an outbound train has been proposed in order to rearrange and line cars in the desirable order. The learning algorithm of the proposed method is derived based on the Q-learning algorithm, considering the total processing time of marshaling. In order to reduce the total processing time of marshaling, the proposed method learns the layout of groups, as well as the arrangement of freight cars in each group, the rearrangement order of cars, the number of cars to be moved and the removal destination of cars, simultaneously. In computer simulations, learning performance of the proposed method has been improved by grouping cars that have the same destination and augmenting the number of candidates of the desired layout.

References

1. G. LI, M. MUTO, N. AIHARA and T. TSUJIMURA, *Quarterly Report of RTRI* **48**, 207 (2007).
2. U. Blasum, M. R. Bussieck, W. Hochstättler, C. Moll, H.-H. Scheel and T. Winter, *Mathematical Methods of Operations Research* **49**, 137 (2000).
3. R. Jacob, P. Marton, J. Maue and M. Nunkesser, Multistage methods for freight train classification, in *Proceedings of 7th Workshop on Algorithmic Approaches for Transportation Modeling, Optimization, and Systems*, 2007.
4. L. Kroon, R. Lentink and A. Schrijver, *Transportation Science* **42**, 436 (2008).
5. N. TOMII and Z. L. Jian, Depot shunting scheduling with combining genetic algorithm and pert, in *Proceedings of 7th International Conference on Computer Aided Design, Manufacture and Operation in the Railway and Other Advanced Mass Transit Systems*, 2000.
6. S. He, R. Song and S. Chaudhry, *European Journal of Operational Research* **124**, 307 (2000).
7. E. Dahlhaus, F. Manne, M. Miller and J. Ryan, Algorithms for combinatorial problems related to train marshalling, in *Proceedings of the 11th Australasian Workshop on Combinatorial algorithms*, 2000.
8. C. Eggermont, C. A. J. Hurkens, M. Modelski and G. J. Woeginger, *Operations Research Letters* **37**, 80 (2009).
9. Y. Hirashima, *IAENG International Journal of Computer Science* **38**, 242 (2011).
10. Y. Hirashima, *Lecture Notes in Engineering and Computer Science: Proceedings of The International MultiConference of Engineers and Computer Scientists 2012, IMECS 2012, 14-16 March, 2012, Hong Kong* , 47 (2012).
11. C. J. C. H. Watkins and P. Dayan, *Machine Learning* **8**, 279 (1992).
12. R. Sutton and A. Barto, *Reinforcement Learning* (MIT Press, 1999).

A WEB-BASED MULTILINGUAL INTELLIGENT TUTOR SYSTEM BASED ON JACKSON'S LEARNING STYLES PROFILER AND EXPERT SYSTEMS

H. MOVAFEGH GHADIRLI

Young Researchers Club, Islamshahr Branch, Islamic Azad University, Islamshahr, Iran
hossein.movafegh@iau-saveh.ac.ir

M. RASTGARPOUR

Department of Computer Engineering, Saveh branch, Islamic Azad University, Saveh, Iran
m.rastgarpour@iau-saveh.ac.ir

Nowadays, Intelligent Tutoring Systems (ITSs) are so regarded in order to improve education quality via new technologies in this area. One of the problems is that the language of ITSs is different from the learner's. It forces the learners to learn the system language. This paper tries to remove this necessity by using an Automatic Translator Component in system structure like Google Translate API. This system carry out a pre-test and post-test by using Expert System and Jackson Model before and after of training a concept. It constantly updates learner model to save all changes in learning process. So this paper offers an E-Learning system which is web-based, intelligent, adaptive, multilingual and remotely accessible where tutors and learners can have non-identical language. It is also applicable Every Time and Every Where (ETEW). Furthermore, it trains the concepts in the best method with any language and low cost.

1. Introduction

In 1982, Sleeman and Brown reviewed the state of the art in computer-aided instruction. They proposed a novel term of ITSs to describe these evolving systems as well as distinguish them from other educational systems. So the implicit assumption corresponding to the learner focused on *learning-by-doing*. They [1] classified available ITSs into four categories include of problem-solving monitors, coaches and teachers, laboratory instructors, and consultants. The first E-Learning software was static and non-intelligent. A course had been organized by a prior procedure and taught in the same style for all learners. They were either computerized versions of textbooks (characterized as electronic page turners) or drill and practice monitors which were giving a learner some problems and comparing the learner's answers with pre-scored answers [1].

There are several kinds of Learners in the Web. It means that some learners need to repeat some lessons mean while some lessons must be removed for others. Later the researchers concluded that learning process must be dynamic and intelligent based on pedagogy view [2]. But developing an applicable and trustful system is so hard [2]. This caused to advent new generation of intelligent educational systems [3].

Web-based learning and *intelligent learning* is so regarded in education todays [2, 4]. A web-based tutor has some benefits like tirelessly, dominance on concepts, low cost and independent of time and place. It can utilizes conversational agents as well [3]. However millions learners of the world can learn via thousands of expert tutor through the web in an intelligent and virtual school. One of the important classes of intelligent tutors is based on rule-based expert system. It determines whatever the student knows, doesn't know and knows incorrectly. It uses this information to adjust learning style. This education style which enjoys the benefits of Expert Systems is called *model tracing tutor* [5].

ITS sometimes has different language of learner language. So such participate needs to know its language in this virtual class. On the other hand, the learning of a foreign language is often difficult and impractical for adults. Moreover, many such courses can be very time consuming, because learners often must cope with unfamiliar writing systems as well as differing cultural norms. So a system is demanded in order to eliminate this "language barrier".

This paper introduces an intelligent system to enjoy Expert Systems abilities. So E-Learning would be efficient, adaptive and just needs a computer and internet connection. Adapting with web-based contexts is very important, because the concept, which is developed for one user, isn't useful for others [2].

The proposed system determines the learning style through a test. It develops a primary model of learner. The learning process starts then. Some characteristics of learner may be varying during learning progress gradually. These progresses would be saved in learner model by system. So learner model gets closer step by step. The system can receives scientific and mental feedback of the learner and change the learning style during the process.

A learner doesn't have to know extra language by using an *Automatic Translation* component. The learner can be read the contents of E-books, virtual blackboard, chat rooms and even write with any known language.

Web-based learning also is useful for training new employees and adapting staffs according to company changes. However proposed system is an artificial environment that easily loses the motivational benefits of authentic task-oriented dialog, but ideally, web-based learning improves the learning efficiency for students and employees through features that are not available in face to face learning. It also allows learners to access the learning materials and interact with the rest of course ETEW. The aim of proposed system is to offer the content which the learner can't be aware of it easily with any language.

The rest of this paper is organized as follows. Section II defines an intelligent tutoring system, automatic translator and presents some available samples. It also deliberates the role of pedagogy in learner modeling Jackson Model. Section III describes the proposed E-Learning system which is intelligent, adaptive, customized, web-based and multilingual. Finally this paper concludes in Section IV.

2. Background

2.1. *Intelligent Tutoring System*

ITSs are computer-based instructional systems with models of educational content. They specify what to teach, and also teaching strategies that specify how to teach [6]. In late 1960, the ITSs have moved out of academic labs and have been applied in classrooms and workplaces. Some of them have shown to be high effective [5]. Unfortunately, intelligent tutors are difficult and expensive to build whereas they are more common and have proven that to be more effective.

Intelligent systems can recognize the learner type, choose appropriate course content from knowledge base and present it to learners in proper style. It also attempts to simulate a human tutor expertly and intelligently. Students using these systems usually solve problems and related sub-problems within a goal space, and receive feedback when their behavior diverges from that of the learner model.

Some design factors have role for each ITS which including component expert simulator, tutor software, learner model, modeler, and knowledge base. Figure 1 illustrates them.

ITSs allow "mixed-initiative" tutorial interactions, where learners can ask questions and have more control over their learning. Some available ITSs are introduced in Table 1.

Some web-based ITSs are presented in Table 2.

2.2. *Automatic Translation*

Automatic Translating System is more efficient than professional human translators in terms of cost and speed of translation. Two types of translation tools which are used frequently are as follows:

- *Machine Translation* is suitable for e-learning localization and multilingual information retrieval.
- *Translation Memories* takes a certain source text and then stores it with corresponding translation done by a human. It can automate the translation process completely in desirable mode.

Machine translation has been appeared in 1981 for personal computers. In 1997, *Babel Fish* appeared the first, free and translation service on the World Wide Web [7].

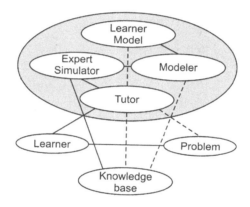

Figure 1. Intelligent Tutoring System Components

Table 1. Some Available Intelligent Tutoring Systems [8]

Name	Comments
CAES system [9]	Has been developed by integration of shipping simulation and intelligent decision system.
ICATS [10]	Coordinates an expert system with multimedia system in an intelligent learning system.
ANDES [11]	Trains Physics without using natural language.
ATLAS [12]	Uses natural language to teach Physics. It can answer to learner's questions.
ELM-ART [13]	A web-based tutoring System for learning programming in LISP.
APHID-2 [14]	A hypermedia generation system with adaption by defining rules that map learner information.
AUTOTUTOR [15]	Be full intelligent and teach *"introduction to computers"* in at least 200 universities of the world.

The *Automatic Translator* includes a Natural Language Parser (NLP). It can annotate phrases with structural information and refer to relevant grammatical explanations. An Error Model which detects and analyzes syntactic mistakes would be referred as well.

The main task of the *Automatic Translator* component is real-time translating of text contents from one language to another. The teacher can store contents with any language in the system without user intervention. It will translate to the learner language then.

Table 2. Adaptive and Intelligent Technologies in Web-based ITS SYSTEMS [16]

System	Hypertext Component	Adaptive Sequencing	Problem Solving Support	Intelligent solution Analysis	Adaptive Presentation
AST	Y	Y			Some
InterBook	Y	Y	Server	Y	Some
PAT-InterBook	Y	Y			
DCG	Y				
PAT	N			Y	
WITS	Y				Y
C-Book	Y	Some			
Manic	Y	Some	Server	Y	Y
Proposed System	Y	Y			Some

It should be noted that multilingual meetings usually is online and also involve more than two learners with different languages. Automatic translation can be applicable in chat among the learners. Table 3 presents some applications which provide automatic translation for instant messaging (Chat).

Table 3. Online Chat Systems with Automatic Translation [17]

Application	Languages
Amikai	9
Annochat	4
ChatTranslator	7
Free2IM Hab.la	13
Realtime Chat	41
IBM Lotus	7
Sametime	7
MeGlobe	15
WorldLingo Chat	15

Any *Automatic Translator* needs to a web-based multilingual dictionary to be useful for online and real-time multilingual translation. Nowadays, the most of web services are provided by online dictionaries supporting two or more languages.

One of the most popular web based dictionary is *Google Translate* to translate each pair of 64 languages to each other such as English to Spanish, French to Russian, Chinese to English, etc. There are 4032 language pairs to translate.

*Google Translate** is based upon a statistical translation system unlike other available machine translation systems. In fact a language model is trained on billions of words out of equivalent text in many different languages. For example, it compares a German word like "Bible" versus a Russian word like "Book". Then it uses comparison results efficiently [18].

The translation of comprehension is still no perfect because the accuracy of Google Translate varies with sentence, vocabulary complexity and by language.

2.3. The Role of Pedagogy in Learner Modeling

During the 1980's, computer scientists especially in AI continued to focus on the problems of natural language, learner models, and deduction.

The *pedagogical strategies* are issued from psychological and didactic research. It pertains with the underlying learning theories, i.e. behaviorism, cognitivism and constructivism. These strategies are specified by a pedagogical specialist – expert in education – and not a computer scientist [18].

The learner's skill profile is recorded in a Learner Model. Jackson models the learner's behavior based on psychology of personality. The Expert modeler monitors learner performance based on Jackson's learning styles profiler.

In fact, pedagogy is a bridge in the gap between learning and modern technologies. It caused ITS helps learner by "*the best way*".

2.4. Jackson's Learning Styles Profiler

Learning styles are various approaches or methods of the learning. An ITS is based on accurate recognition of behaviors and individual characteristics. There are some important factors in learning style such as aptitude, personality and behavior [11].

---

* http://translate.google.com/

It's worth to be noted that the concept of "learning style" has emerged as a focal point of much psychological research. In our proposed learner model, we have adopted the Jackson's learning styles profiler in order to model learning styles of the learners. Jackson proposed five learning styles which are summarized in Table 4. Reader can find more details in [2].

Table 4. Summarization of Learning Styles [8]

Learning Style	Specifications
Sensation Seeking (SS)	Believing that the experiences create learning.
Goal Oriented Achievers (GOA)	Self-confident to achieve difficult and certain target.
Emotionally Intelligent Achievers (EIA)	Rational and goal-oriented.
Conscientious Achievers (CA)	Responsible and insight creator.
Deep Learning Achievers (DLA)	Interested in learning highly.

There are some approaches to model learner's behavior. For example Entistle approach [21] tries to connect the psychology concepts into the effective variables on the learner's view to *'learning'*. They are rarely applied in adaptive education systems despite some benefits in education and psychology of personality.

This section introduces Jackson model [22]. Jackson's model is based on the most recent researches in the psychology of personality. It also is an efficient analyzer. So this paper just applies it among available models.

Jackson model has been proposed in Queensland University [22]. This model has been developed by investigating personality, learning and evaluating for fifteen years. The learning is based on new neurological psychology in this model.

This model has some advantages. For example, it can export and report the individual specifications of the learner. It is obtained on basis of the points which the learner earns in the learning style (III-D).

The aim of this model is to recognize an ideal style based on pedagogy principles which can model learner's behavior and specifications in adaptive E-Learning system. A key premise of Jackson's model is that cognitive strategies redirect Sensation Seeking to predict functional behaviors [8].

3. Proposed Approach

ITSs build a model of the individual learner's knowledge, difficulties and misconceptions during interaction with the system. This learner model can be compared with a model of the target domain. Suitable tutorial strategies can be inferred by the system. They are appropriate for the learner according to the contents of their learner model. In other words the educational interaction will be matched with specific and unique needs of each learner [23].

A web-based ITSs is an E-Learning system based on web which can be used ETEW. The first E-Learning system has been reported on 1995 [17, 24]. It is web-based and intelligent. Learning of all courses is customized well at home via web in this system. So learners can solve some examples and proper exercises ETEW. Finally he can take a course exam virtually or physically.

A large number of ITSs have been conducted in a single language so far. However, ITSs can support multiple languages with the integration of machine translation such as *Google Translate API* among 64 languages.

3.1. *Learning Environment*

The learning environment is explained in this section. A learner can visits website of intelligent virtual school after authentication. An intelligent Graphical User Interface (GUI) is an interface between learners and intelligent tutor. This section of system can affect learning efficiency. It should be user friendly as far as possible.

An intelligent virtual class has some benefits such as graphical properties, audio and video to make learning attractive. Moreover, some tools are available to simplify learning process. Learners can communicate well with this inanimate and non-physical system via these tools. Some facilities are Computer games, frequently asked questions (FAQ), Video chat and email.

Various learners may be logged with different language according to geographic region. Moreover all of E-learning systems can be customized for different languages. For example, the same e-learning system can be run for Iranian people with Persian contents.

This system has an *Automatic Translator* component which is able to integrate language translatability with the system. Language modeling for each learner is performed and store in the learner model. This component is developed with a MultiTranslator such Google Translate API that provides free, cheap and true translation in many languages at the same time.

Translation cost [25] and grammatical errors are some of the major limitations in localizing e-learning system. They are reduced along with growing of these systems.

This component helps learners to use the system without knowing any foreign language. Some users can also improve their skills in second language.

3.2. *Learning Process and Evaluation*

This system uses a three layered structure to implement a concept: *"Pre-test"*, *"Learning concept"* and *"Post-test"*.

The pre-test includes of some questions planned by an expert tutor in advance. These questions can determine learner's primary knowledge level. It simulates learner model based on Jackson model [22]. The learning concept depends on the learner level. So the best method is determined to train a learner. A learning process starts up then. A post-test will evaluate the learner by some questions finally.

Learning evaluation is the most important factor to determine learning performance in E-Learning systems. It is performed with *Pre-test* and *Post-test* layers of the proposed system. The evaluation has two levels, *conceptual* and *objective*. The first one refers to the learner's understanding out of the lesson *concept*. The last one denotes to the learner's understanding of the lesson *topic*. The learner's knowledge level is determined by *concept level* and *objective level*.

The tutor can extract proper questions from question-base collected by an *expert system*, pre-test and post-test. He notes that a specific score is given to each question.

Question selection should satisfy some rules:

1. None of them should be repetitious even if a learner would be trained a specific concept several times.
2. The question must be planned for all sections of a concept entirely.
3. An expert tutor plans questions in all level.

The sequence, number and level of questions are determined according to learner level and learning type intelligently. Sum of scores is calculated and then learner level is determined after answering the questions. Table 5 presents five categories of learner's knowledge level about a concept [13, 14].

Table 5. Categories of Knowledge Level [8]

Knowledge Level	Score
Excellent	86-100
Very good	71-85
Good	51-70
Average	31-50

The system modeler updates the learner's model along with questions answering. It can also save last academic status of learner and all his learning records.

The proposed system integrates *pedagogy* and *expert system* with web-based ITS structures. It develops a multilingual e-learning system by applying automatic translator algorithms which is illustrated in Figure 2. So it can be used by different nationalities. In addition of education application, it can be used as a tool to learn a second language.

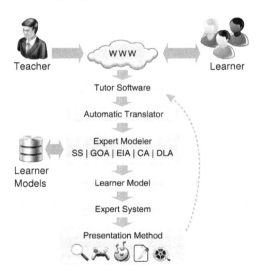

Figure 2. Block Diagram of Proposed System

4. Conclusion

This paper proposed a model for an adaptive, intelligent, multilingual and web-based tutor. This model can models the learning styles of the learners using Jackson's learning styles profiler and expert system technology to enhance learning performance. Previous E-Learning systems offer fixed languages and multimedia web pages and facilities for user management and communication.

Based on the proposed model, E-Learning software is implemented on the web which can identify learning styles, aptitude, characteristics and behaviors of learner in order to provide appropriate individual learning content. On the other hand, Jackson model can recognizes learner's learning style. It leads to make learner's model closer. So the efficiency of adaptation process increases.

The proposed model acts as an intelligent tutor which can perform three processes include of *pre-test*, *learning concept* and *post-test* based on learner's characteristic. This system uses expert simulator and its knowledge base as well.

Proposed model is also *multilingual.* So it helps learners to use the system without knowing any foreign language. Some users can improve their second language skills by this way.

A large number of ITS have been ever conducted in a single language. By using the proposed model, ITSs can support multiple languages with the integration of machine translation such as *Google Translate API* among 64 languages and 4032 language pairs.

The proposed system doesn't have any drawback of previous system and human expert tutor. It can improve the learning result significantly. In other words, it helps learners to study in *"the best way"* using localized language.

References

1. K. Eustace, "Educational value of E-Learning in conventional and complementary computing education," in Proceedings of the 16th National Advisory Committee on Computing Qualifications (NACCQ), Palmerston North, New Zealand, 2003, pp. 53–62.
2. H. M. Ghadirli and M. Rastgarpour, "A Web-based Adaptive and Intelligent Tutor by Expert Systems," in The *Second International Conference on Advances in Computing and Information Technology (ACITY 2012)*, Chennai, India, 2012.
3. P. Brusilovsky, "Intelligent tutoring systems for World-Wide Web," in *Third International WWW Conference(Posters)*, Darmstadt, 1995, pp.42-45.
4. M. a. O. Specht, R., "ACE-Adaptive Courseware Environment," *The New Review of Hypermedia and Multimedia*, vol. 4, pp. 141-161, 1998.
5. P. Brusilovsky, "Methods and techniques of adaptive hypermedia," *User Modeling and User-Adapted Interaction*, vol. 6, pp. 87-129, 1996.
6. S. Ohlsson, "Some Principles of Intelligent Tutoring," *In Lawler & Yazdani (Eds.), Artificial Intelligence and Education*, vol. 1, Ablex: Norwood, NJ, pp. 203-238, 1987.

7. J. Yang and E. Lange, "SYSTRAN on AltaVista: A user study on real-time machine translation on the Internet," *Proceedings of the 3rd Conference of the Association for Machine Translation in the Americas*, Langhorne, pp. 275-285, 1998.

8. H. M. Ghadirli and M. Rastgarpour, "A Model for an Intelligent and Adaptive Tutor based on Web by Jackson's Learning Styles Profiler and Expert Systems," Lecture Notes in Engineering and Computer Science: *Proceedings of The International MultiConference of Engineers and Computer Scientists 2012*, IMECS 2012, 14-16 March, 2012, Hong Kong, pp 63-67.

9. E. Shaw, *et al.*, "Pedagogical agents on the web," in *Proceedings Of Third International Conference on Autonomous Agents*, 1999, pp. 283-290.

10. Shute, V.J. and Regian, J.W. (1990). "Rose Garden Promises of Intelligent Tutoring Systems: Blossom or Thorn?," Presented at Space Operations, Automation and Robotics Conference, June 1990, Albuquerque, NM.

11. Koedinger, K., & Anderson, J. (1995). "Intelligent tutoring goes to the big city," *Proceedings of the International Conference on Artificial Intelligence in Education*, Jim Greer (Ed). AACE: Charlottesville, VA pp. 421-428.

12. C. Jackson, *Learning Styles and its measurement: An applied neuropsychological model of learning for business and education*: UK: PSi-Press, 2002.

13. Weber, G., & Brusilovsky, P. (2001). "ELM-ART an adaptive versatile system for web-based instruction," *International Journal of Artificial Intelligence and Education*, this volume.

14. L. Kettel, J. Thomson, and J. Greer, "Generating individualized hypermedia applications," In *Proceedings of ITS-2000 workshop on adaptive and intelligent webbased education systems*, 2000.

15. N. Entwistle, *Styles of learning and teaching: an integrated outline of educational psychology for students, teachers, and lecturers*: David Fulton Publish, 1989.

16. P. Brusilovsky, "Adaptive and intelligent technologies for web-based eduction," *KI*, vol. 13, pp. 19-25, 1999.

17. Bra PD, Santic T, Brusilovsky P. AHA! meets Interbook, and more. *Proceedings of the World Conference on E-Learning*, Phoenix, AZ, November 2003, Rossett A (ed.). AACE: Norfolk, VA, 2003; 57–64.

18. M. Aiken, *et al.*, "Automatic translation in multilingual electronic meetings, " *Translation Journal*, vol. 13, 2009.

19. M. Siadaty and F. Taghiyareh, "PALS2: Pedagogically adaptive learning system based on learning styles, " 2007, pp. 616-618.

20. C. Yang, "An Expert System for Collision Avoidance and Its Application," PH.D. Thesis, 1995.

21. J. M. Ragusa, "The Synergistic Integration of Expert Systems and Multimedia within an Intelligent Computer Aided environmental tutoring system," in *Proceedings of the 3rd world congress on expert systems*, 1960.

22. A. S. Gertner, VanLehn, K., "ANDES: A Coached Problem-Solving Environment for Physics. In Intelligent Tutoring Systems," in *Fifth International Conference*, ITS New York, 2000, pp. 133–142.

23. Y. Cui and S. Bull, "Context and learner modelling for the mobile foreign language learner," *System*, vol. 33, pp. 353-367, 2005.

24. K. F. VanLehn, R.; Jordan, P.; Murray, C.; Osan, R.Ringenberg, M.; Rosé, C. P.; Schulze, K.; Shelby, R.; Treacy, D.; Weinstein, A.; and Wintersgill,, "Fading and Deepening: The Next Steps for ANDES and Other Model-Tracing Tutors," *Intelligent Tutoring Systems*, 2000.

25. B. L. Hafsa and E. M. des Ingénieurs, "E-learning globalization in multilingual and multicultural environment," *WSEAS Proceedings NNA-FSFS-EC*, pp. 29-31, 2003.

AUTOMATIC MEDICAL IMAGE SEGMENTATION BY INTEGRATING KFCM CLUSTERING AND LEVEL SET BASED FTC MODEL

M. RASTGARPOUR

Department of Computer Engineering, Faculty of Engineering, Saveh branch, Islamic Azad University, Saveh, Iran
m.rastgarpour@iau-saveh.ac.ir;m.rastgarpour@gmail.com

J. SHANBEHZADEH

Department of Computer Engineering, Kharazmi University, Tehran, Iran
jamshid@tmu.ac.ir;shanbehzadeh@gmail.com

Recently, researchers integrate available methods in order to automate medical image segmentation, resolve their drawbacks and enjoy their benefits. Geometric deformable models based on level set have shown promising results in this area. Because, they can capture objects topologies and automatically adapt to topological changes. Among all extensions of them, Fast Two Cycle (FTC) model is the fastest one while retaining significant accuracy. But contour initialization affects its efficiency extremely. This paper proposes an integrative segmentation approach including two successive stages as follows. Firstly, the KFCM clusters input image. Then ROI's fuzzy membership matrix is injected to next stage as an initial contour. Ultimately, FTC model segments the image by curve evolution based on level set. This approach has valuable benefits including automation, noise invariant and high efficiency in terms of accuracy and speed. Simulation results show promising outputs in segmentation of different modalities of medical images.

1. Introduction

Quantitative analysis of medical images, namely volume measurement, needs to determine anatomy structures provided by segmentation [1, 2]. Moreover, segmentation is so regarded for more concentration in later steps in medical image analysis such as feature extraction, measurement and representation of Region of Interest (ROI). So segmentation is a crucial step in image analysis and affects the efficiency of image analysis extremely [3]. Furthermore, as proper diagnosis is so regarded in medical problems, segmentation is very important step in medical image analysis [4].

Some factors complicate segmentation in medical images such as normal anatomic variation, post-surgical anatomic variation, vague and incomplete boundaries, inadequate contrast, inhomogeneity of object boundary, noise and motion blurring artifacts and so on [5]. These factors lead to develop a plenty of researches and several approaches in medical image segmentation. Moreover, some tools are available in this area [6]. Nevertheless, it has been remained a challenging area yet where there is a highly growing interest in this field [7].

Many methods have been proposed for medical image segmentation over the past 30 years [8]. All of them are successful just in a special problem and data. Each method has drawbacks and benefits so that they encourage researchers to integrate them. In this way, not only they can get the benefits but also their disadvantages may be reduced. Moreover, there is no need to initialize the contour by user. So automation and reduction in user interaction have been provided.

So far, several combination methods have been proposed by researchers for automatic medical image segmentation. For example *Watersnake* which has been proposed by integration of *watershed* and *parametric active contours*. Wu et al [9] integrated *KFCM clustering* [10, 11] by *Active contours without edge (CV model)* [12] for MRI brain segmentation. Reddy et al [13] concatenated *KFCM clustering* [10, 11] with an extension of level set based method entitled *Evolution without re-initialization* [14] in order to segment noisy medical images.

Among available segmentation methods, Level set-based geometric active contours [15-17] by theory of curve evolution shows promising results in medical images [18]. Because it can capture the topology of shapes and are robust in noisy images. But this method [16, 17] has some drawbacks. Some researches proposed several extensions to solve them. Consequently, variant level set based algorithms were formed. Each algorithm can solve a part of drawbacks but cause some other problems. Table 1 summarizes the problems of primary version [16, 17] along with corresponding proposed method to solve them.

Table 2 investigates these level set algorithms in terms of general properties like the type of energy and evolution, accuracy based on Dice coefficient [24] and speed by running on the cardiac MRI image [25]. As table 2 shows, *FTC model with Smoothness Regularization* which has been proposed by Shi and Karl [23], is the fastest method among level set based ones while retaining significant accuracy. So it can be called as the most efficient level set based method for medical image segmentation which has ever been proposed.

Table 1 Problems of primary version [16, 17] and corresponding proposed method to solve it

Problem	Proposed method
Failure in ambiguous and discrete edges	Geodesic Active Contour [19]
Noise sensitivity	CV model [12]
Dependence on curve initialization	Variational B-Spline Level set [20]
Re-initialization of sign distance function	Evolution without re-initialization [14]
Unsuccessful to find interior object contour	Variational B-Spline Level set [20] and Region-Scalable Fitting Energy [21]
Failure in inhomogeneous region	Region-Scalable Fitting Energy [21], Localizing Region-Based Active Contours [22]
Computational complexity	FTC Model [23]

Table 2 Investigation of level-set based algorithms by energy, evolution, speed and accuracy

Method	Energy	Evolution	Run-time (s)	Accuracy
Geodesic Active Contour [19]	Contour-based	Narrow band	9.42	0.96
CV model [12]	Region-based	Narrow band	0.77	0.96
Variational B-Spline Level set [20]	Region-based	Total Image	1.12	0.52
Region-Scalable Fitting Energy [21]	Localized Region-based	Total Image	33.15	0.46
Localizing Region-Based Active Contours[22]	Localized Region-based	Narrow band	28.45	0.97
FTC Model[23]	Region-based	Narrow band	0.42	0.96

Unlike desirable results of FTC model, its efficiency is dependent on curve initialization very much. The reason is that initial curve doesn't consider any topological constraints. It also should enclose ROI and touch every object and background region. Figure 1 shows this problem in segmentation of a CT image of vessels. From left to right, an improper curve initialization in the first image fails in segmentation shown in second image. While a proper one in the third image results in accurate segmentation in forth image.

The authors could overcome initial curve dependency by using Kernel Fuzzy C-Mean Clustering algorithm [11] before in previous version of this paper [26]. As that paper has shown promising result on several types of image, now this paper tries to enjoy its benefits in medical image segmentation.

The rest of paper is organized as follows. The basic concepts of KFCM clustering and FTC model are reviewed in section 2. The proposed approach is explained in section 3. Section 4 presents experimental results and discusses advantages and disadvantages of proposed approach. Section 5 mentions the future research and work. This paper will conclude in section 6 finally.

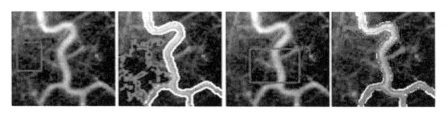

Figure 1 Dependency of FTC Model on curve initialization. Left to right– first and third: curve initialization in red; second and forth: results of segmentation in magenta and reference contours in white.

2. Background

This paper uses an integrative approach including KFCM clustering and FTC model for medical image segmentation. So the basic information about KFCM clustering method and FTC Model is explained in the following.

2.1. *KFCM Clustering*

A Kernel-Based Fuzzy C-Means clustering (KFCM) algorithm has been proposed by Zhang et al [10, 11] with strong noise robustness. In fact, it is obtained just by replacing a new kernel-based metric in the original Euclidean norm metric of FCM.

The KFCM partitions a dataset $X = \{x_1, x_2, ..., x_n\} \subset R^P$, where P the dimension, into c fuzzy subsets by minimizing the following objective function:

$$J_m(U,V) = \sum_{i=1}^{c} \sum_{k=1}^{n} u_{ik}^m \|\Phi(x_k) - \Phi(v_i)\|^2 \tag{1}$$

where c is the number of clusters and determined by a prior knowledge , i.e. c=4 for brain image; n is the number of data points; u_{ik} is the fuzzy membership of x_k in class i; m is a weighting exponent on each fuzzy membership and controls clustering fuzziness (usually $m = 2$); V is the set of cluster centers or prototypes $v_i \in R^P$;and Φ is an implicit nonlinear map. It's better to mention that u_{ik} is a member of [0,1] and must satisfy the following conditions:

$$\sum_{i=1}^{c} u_{ik} = 1 \tag{2}$$

$$0 < \sum_{k=1}^{n} u_{ik} < n \tag{3}$$

In feature space, a kernel can be a function which is called K, where $K(x, y) = \langle \Phi(x), \Phi(y) \rangle$ and $\langle . \rangle$ is the inner product. So:

$$\begin{aligned}
\|\Phi(x_k) - \Phi(v_i)\|^2 &= \left(\Phi(x_k) - \Phi(v_i)\right)^T \left(\Phi(x_k) - \Phi(v_i)\right) \\
&= \Phi(x_k)^T \Phi(x_k) - \Phi(v_i)^T \Phi(x_k) - \Phi(x_k)^T \Phi(v_i) \\
&\quad + \Phi(v_i)^T \Phi(v_i) \\
&= K(x_k, x_k) + K(v_i, v_i) - 2K(x_k, v_i)
\end{aligned} \tag{4}$$

There are some popular kernel functions in [27]. In this paper, the authors use *Gaussian Radial Basis Function* (GRBF) kernel with following formula:

$$K(x, y) = \exp\left(\frac{-\|x - y\|^2}{\sigma^2}\right) \tag{5}$$

where σ is the width parameter, then:

$$K(x, x) = 1 \tag{6}$$

By substituting Eq. (2) and Eq. (3) in Eq. (1), we have:

$$J_m(U, V) = 2 \sum_{i=1}^{c} \sum_{k=1}^{n} u_{ik}^m [1 - K(x_k, v_i)] \tag{7}$$

Similar to FCM, the optimization problem comes to minimize $J_m(U, V)$ under the constraints of u_{ik}. Then:

$$u_{ik} = \frac{(1 - K(x_k, v_i))^{-1/(m-1)}}{\sum_{j=1}^{c}(1 - K(x_k, v_j))^{-1/(m-1)}} \tag{8}$$

and:

$$v_i = \frac{\sum_{k=1}^{N} u_{ik}^m K(x_k, v_i) x_k}{\sum_{k=1}^{N} u_{ik}^m K(x_k, v_i)} \tag{9}$$

Summarization of KFCM algorithm is described in Table 3. In this algorithm, similarity measure in FCM, i.e. Euclidean norm metric, is replaced by a new kernel-induced metric (in this paper, Gaussian kernel) which makes the weighted sum of data points more robust (for more information refer to [11]). So this algorithm is a robust clustering approach if an appropriate value for sigma would be chosen. The sigma value is obtained by trial-and-error technique or experience or prior knowledge which neither too large nor too small. The authors consider 150 for sigma like [11] does.

As mentioned in [11], KFCM needs to be improved for medical image segmentation. This paper applies the power of curve evolution based on level set to increase the efficiency of segmentation by KFCM.

Table 3 Algorithm of KFCM clustering

I. Initialize c, t_{max}, m>1, $\varepsilon > 0$ for positive constants.

II. Initialize the membership matrix u_{ik}^0

III. For t=1 to t_{max} do:

    a) Update all prototype v_i^t with $v_i = \frac{\sum_{k=1}^N u_{ik}^m K(x_k, v_i) x_k}{\sum_{k=1}^N u_{ik}^m K(x_k, v_i)}$

    b) Update all memberships u_{ik}^t with $u_{ik} = \frac{(1-K(x_k, v_i))^{-1/(m-1)}}{\sum_{j=1}^c (1-K(x_k, v_j))^{-1/(m-1)}}$

    c) Compute $E^t = max_{i,k}|u_{ik}^t - u_{ik}^{t-1}|$, if $E^t \leq \varepsilon$, stop;
    End;

2.2. Fast Two Cycle Model

Shi and Karl [23] approximated the level-set function in a narrow band using an integer valued array. To speed up the computations for real time application, they used a Fast Two Cycle (FTC) algorithm instead of solving Partial Differential Equations (PDEs). The basic idea is to use two cycles for curve evolution, one cycle for data-dependent term and another one for smoothness regularization. Implicit function is approximated by a piece wise constant function and defined by only four values for fast computation:

$$\phi(x) = \begin{cases} 3 & x \text{ is extrior point} \\ 1 & x \in L_{out} \\ -1 & x \in L_{in} \\ -3 & x \text{ is interior poin} \end{cases} \qquad (10)$$

where

$$\begin{cases} L_{out} = \{x | x \in \Omega \ \ and \ \exists y \in N(x); y \in D \backslash \Omega\} \\ L_{in} = \{x | x \in D \backslash \Omega \ \ and \ \exists y \in N(x); y \in \Omega\} \\ \qquad N(x) = \left\{ y \in D \ \middle| \ \sum_{k=1}^{K} |y_k - x_k| = 1 \right\} \end{cases} \qquad (11)$$

Consider C is a curve which can evolve iteratively under a speed function and stop when it converges. The set of grid points enclosed by C as the object region Ω and the set of points $D \backslash \Omega$ as the background region. The interior points are those grid points inside C but not in L_{in} and the exterior points are those points outside C but not in L_{out} similarly. This function approximates the signed distance function locally.

Their algorithm can apply some evolution speeds which composed of a data dependent term and a curve smoothness regularization term. So the general speed function F is $F = F_{int} + F_d$. In this paper, the authors use speed function of *CV model* [12] for data-dependent speed. So:

$$F_d = H(\phi(x))(I(x) - v)^2 + (1 - H(\phi(x)))(I(x) - u)^2 \qquad (12)$$

where I is the intensity, H is Heaviside function defined as:

$$H(z) = \begin{cases} 1 & z \geq 0 \\ 0 & z < 0 \end{cases} \qquad (13)$$

$$u = \frac{\int_{\Omega} \left(1 - H(\varphi(x))\right) . I(x) dx}{\int_{\Omega} \left(1 - H(\varphi(x))\right) dx} \qquad (14)$$

and

$$v = \frac{\int_{\Omega} H(\varphi(x)) . I(x) dx}{\int_{\Omega} H(\varphi(x)) dx} \qquad (15)$$

Then they just use the sign of evolution speed function F. So,

$$\widehat{F_d} = \begin{cases} 1 & F_d > 0 \\ 0 & F_d = 0 \\ -1 & F_d < 0 \end{cases} \qquad (16)$$

They use MBO algorithm [28, 29] for smoothing speed, F_{int}, which is derived Gaussian filtering:

$$\hat{F}_{int} = \begin{cases} 1 & for \; x \in L_{out} \, , if \; G \otimes H(-\phi)(x) > \dfrac{1}{2} \\ -1 & for \; x \in L_{in} \, , if \; G \otimes H(-\phi)(x) < \dfrac{1}{2} \\ 0 & otherwise \end{cases} \tag{17}$$

where $\otimes$ is the convolution operation.

This algorithm is fast enough to be applied in real-time application. The cause is that the evolution steps use only integer operations (see Eq. 10) in both cycles by two simple element switching mechanisms between two linked lists. They called these mechanisms, *switch_in* to move the boundary outward by one grid point at that location and similarly *switch_out* to move the boundary inward. The algorithms of switching mechanisms and FTC are represented in Table 4 and Table 5 respectively.

Table 4 *Switch_in* and *Switch_out* algorithms

Switch_in(x)	Switch_out(x)
For a point $x \in L_{out}$ 1. Delete x from L$_{out}$. 2. Add those N(x) which were exterior points to L$_{out}$. 3. Insert x to L$_{in}$.	For a point $x \in L_{in}$ 1. Delete x from L$_{in}$. 2. Add those N(x) which were interior points to L$_{in}$. 3. Insert x to L$_{out}$.

3. Proposed Approach

As mentioned in *Introduction* section, some researchers have concatenated several methods to automate medical image segmentation. In other hand, as proved there, the efficiency of FTC model (one of the most efficient algorithm in this area) depends on curve initialization extremely. Because of curve evolves just in a narrow band of initial curve. This paper uses an integrative approach in which the KFCM clustering is applied in order to improve and automate medical image segmentation.

The proposed approach consists of two successive stages to integrate FTC model with KFCM clustering. Firstly, the KFCM is used to extract ROI's fuzzy membership matrix. Secondly, zero level set is initialized using this matrix and curve would be evolved iteratively by FTC model to converge and segment the ROIs. Figure 2 illustrates the framework of proposed approach. The algorithm of proposed method can be seen in Table 6 as well.

Table 5 FTC Algorithm for image segmentation by curve evolution based on level set [23]

1. **Initialize** *following values:* *a)* ϕ, $\widehat{F_d}$ and $\hat{F}_{int}$ *based on initial curve.* *b)* t_{max} (*predefined maximum iterations)*

2. **Cycle 1: data dependent evolution**
For i=1 to N_a *do*
a) Compute F_d *for each point in* L_{out} *and* L_{in} *(in Eq. 11)and store its sign in* $\widehat{F_d}$*, so:*
$$\widehat{F_d} = \begin{cases} 1 & F > 0 \\ 0 & F = 0 \\ -1 & F < 0 \end{cases}$$
b) Outward evolution: For each point $x \in L_{out}$, *if* $\widehat{F_d}(x) > 0$: *switch_in(x) (see Table 4).*
c) Eliminate redundant points in L_{in} *(in Eq. 11).*
    For each point $x \in L_{in}$, *if* $\forall y \in N(x)$ *(in Eq. 11),* $\varphi^\gamma(y) < 0$:
    i. Delete x *from* L_{in}
    ii. Set $\varphi^\gamma(x) = -3$.
d) Inward evolution. For each point $x \in L_{in}$, *if* $\widehat{F_d}(x) < 0$: *switch_out(x) (see Table 4).*
e) Eliminate redundant points in L_{out}. *For each point* $x \in L_{out}$, *if* $\forall y \in N(x)$, $\varphi^\gamma(y) > 0$,
    i. Delete x *from* L_{out}
    ii. Set $\varphi^\gamma(x) = 3$.
f) Check following stopping condition, if it is satisfied, go to 3; otherwise continue this cycle.
    I. The speed at all the neighboring grid points satisfies:
$$\hat{F}(x) = \begin{cases} \leq 0 & ; for\ all\ x \in L_{out} \\ \geq 0 & ; for\ all\ x \in L_{out} \end{cases}$$
    II. $T = t_{max}$
End;

3. **Cycle 2: smoothing via Gaussian filtering**
For i=1 to N_s *do*
a) Compute the smoothing speed $\hat{F}_{int}$ *for each point in* L_{out} *and* L_{in}.
b) Outward evolution. For each point $x \in L_{out}$, *if* $\hat{F}_{int}(x) > 0$: *switch_in(x).*
c) Eliminate redundant points in L_{in}. *For each point* $x \in L_{in}$, *if* $\forall y \in N(x)$, $\varphi^\gamma(y) < 0$:
    i. Delete x *from* L_{in}
    ii. Set $\varphi^\gamma(x) = -3$.
d) Inward evolution. For each point $x \in L_{in}$,*) if* $\hat{F}_{int}(x) < 0$: *switch_out(x).*
e) Eliminate redundant points in L_{out}. *For each point* $x \in L_{out}$, *if* $\forall y \in N(x)$, $\varphi^\gamma(y) >$:
    i. Delete x *from* L_{out}
    ii. Set $\varphi^\gamma(x) = 3$.
    End;

4. *If* **stopping condition** *not satisfied in 2, go to 2.*

Table 6 The algorithm of **proposed method** (integration of KFCM and FTC) [26]

1. **Stage 1: the KFCM clustering**
   i. Set value of c, t_{max} *(maximum iteration number),* $m=2$, $\sigma = 150$, $\varepsilon = 0.001$.
   ii. Continue from step 2 in Table 3 which consists of KFCM algorithm.
   iii. Extract ROI's fuzzy membership matrix.

2. **Stage 2: Curve evolution by FTC model**
   i. Initialize the zero level set with the output matrix of stage 1(step iii) with following function
   [9, 30] : $\phi(x, y, t = 0) = \begin{cases} 1 & inside\ ROI \\ 0 & outside\ ROI \end{cases}$
   ii. Continue from step 1 in Table 5 (the algorithm of FTC Model).

Figure 2 Framework of proposed approach

4. Experimental Results

To evaluate proposed approach, the authors applied different kinds of medical images. They implemented the proposed approach (section 3) in *Matlab* version R2008a by a 2.27 GHZ Intel Core (TM) 2 Duo CPU processor with 2 GB RAM.

In implementation of KFCM, the parameters m and σ are fixed for all input images while the number of clusters, parameter c, varies with respect to input image. It is determined based on prior knowledge of image. It was set value "3" for the images of vessel and retina; and value "4" for brain images in this simulation.

The algorithm of FTC model (table 5) has four parameters, N_a, N_s, N_g and σ. N_a controls data-dependent speed and the others effects on smoothing regulation speed. Their experiments showed that FTC model is robust with respect to perturbation of these parameters [23]. So these parameters are generally chosen as follows: $N_a=30$, $N_s=3$, $N_g=3$ and $\sigma = 1$ similar to those they applied. The max iteration, t_{max}, was considered 200 as well.

Figure 3 depicts the simulation results of proposed approach with the following details of input images. Vessels CT image has been borrowed from [21], brain MRI image is from [31], retina vessels tortuosity has been copied from [32] and finally brain MRI image is related to [33] from top to bottom respectively. The columns are original image, the result of KFCM clustering, and the result of integrating KFCM with FTC model (proposed approach) from left to right respectively.

The second column of figure 3 shows that the KFCM clustering can't operate well alone. Moreover FTC model is high dependent on initial curve as depicted on figure 1. The third column of figure 3 illustrates promising result of this integrating in red for ROIs boundaries.

4.1. *Evaluation of Proposed Approach*

This section evaluates the proposed approach in terms of advantages and disadvantages. It has valuable benefits such as automation, high accuracy and speed, independent on initial condition, noise invariant and success in poor boundaries with high geometric complexity.

As mentioned before, the problem of FTC is dependency on initial contour. Since it has been resolved by using KFCM, it should be noted that drawbacks of the proposed method are related to KFCM clustering. They include the need to prior knowledge about the number of clusters, using experimental and approximate sigma, lack of using spatial information of pixels and failure in inhomogeneous objects.

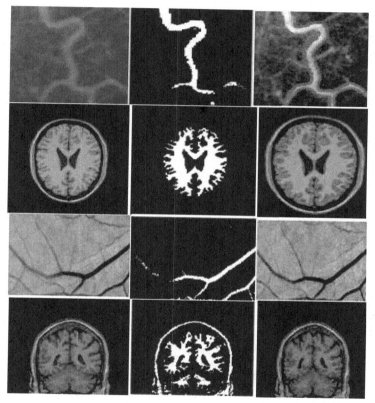

Figure 3 Simulation results for variant medical images. The columns from left to right: original images, KFCM clustering results and proposed approach results (in red).

5. Future Research

In the future, the authors are going to compare the proposed approach with similar approaches, i.e. [9, 13] in terms of speed. Some applied parameters in this paper like sigma have been *empiric*. Hence, the authors would like to determine them based on input data later.

This method should be examined in noisy conditions. For this purpose, smoothness regularization will be considered later. Furthermore, resolving the drawbacks in section 4.1 is another future work.

6. Conclusion

This paper presented an integrative approach for automatic medical image segmentation. It derived the benefit of Curve evolution by FTC model and the KFCM clustering by integrating them. In fact, FTC model is so efficient in terms of computational complexity and accuracy among available methods in this area. But curve initialization affects its efficiency extremely. This paper used fuzzy membership matrix of KFCM for initial curve to solve this problem. The proposed approach had valuable benefits such as automation, high accuracy and speed, independent on initial condition, noise invariant and success in poor boundaries with high geometric complexity. Simulation result showed promising outputs for segmentation of variant medical images including vessels CT, brain MRI and retina vessels tortuosity.

References

1. Isgum, I., et al., *Multi-atlas-based segmentation with local decision fusion—Application to cardiac and aortic segmentation in CT scans.* Medical Imaging, IEEE Transactions on, 2009. **28**(7): p. 1000-1010.
2. Rastgarpour, M. and J. Shanbehzadeh. *Novel Classification of Current Methods, Available Softwares and Datasets in Medical Image Segmentation.* in *The 2011 World Congress in Computer Science, Computer Engineering, and Applied Computing (WORLDCOMP'11).* 2011. LV, USA.
3. Gonzalez, R.C. and E. Richard, *Woods, digital image processing.* 2002, Prentice Hall Press, ISBN 0-201-18075-8.
4. Rastgarpour, M. and J. Shanbehzadeh, *The Status Quo of Artificial Intelligence Methods in Automatic Medical Image Segmentation.* International Journal of Computer Theory and Engineering (IJCTE) 2012: p. -in press.
5. Wang, S. and R.M. Summers, *Machine Learning and Radiology.* Medical Image Analysis, 2012. **16** (5): p. 933-951.
6. Rastgarpour, M. and J. Shanbehzadeh, *Application of AI Techniques in Medical Image Segmentation and Novel Categorization of Available Methods and Tools.* Lecture Notes in Engineering and Computer Science: Proceedings of The International MultiConference of Engineers and Computer Scientists 2011, IMECS 2011, 2011. **2188**: p. 519-523.

7. Rastgarpour, M. and J. Shanbehzadeh, *The Problems, Applications and Growing Interest in Automatic Segmentation of Medical Images from the year 2000 till 2011.* International Journal of Computer Theory and Engineering (IJCTE) 2012: p. -in press.

8. Zhang, Y.J., *Advances in image and video segmentation.* 2006: Irm Pr.

9. Wu, Y., W. Hou, and S. Wu, *Brain MRI segmentation using KFCM and Chan-Vese model.* Transactions of Tianjin University, Springer, 2011. **17**(3): p. 215-219.

10. Zhang, D.Q., et al. *Kernel-based fuzzy clustering incorporating spatial constraints for image segmentation.* in *Proc. of the 2th Int. Conf. on Mach. Lear. and Cyb.* 2003: IEEE.

11. Zhang, D.Q. and S.C. Chen, *A novel kernelized fuzzy c-means algorithm with application in medical image segmentation.* Artificial Intelligence in Medicine, 2004. **32**(1): p. 37-50.

12. Chan, T.F. and L.A. Vese, *Active contours without edges.* Image Processing, IEEE Transactions on, 2001. **10**(2): p. 266-277.

13. Reddy, G., et al. *Image segmentation using kernel fuzzy c-means clustering on level set method on noisy images.* in *International Conference on Communications and Signal Processing (ICCSP).* 2011: IEEE.

14. Li, C., et al. *Level set evolution without re-initialization: a new variational formulation.* in *Computer Vision and Pattern Recognition (CVPR) IEEE Computer Society Conference on.* 2005.

15. Osher, S. and J.A. Sethian, *Fronts propagating with curvature-dependent speed: algorithms based on Hamilton-Jacobi formulations.* Journal of computational physics, 1988. **79**(1): p. 12-49.

16. Caselles, V., et al., *A geometric model for active contours in image processing.* Numerische Mathematik, 1993. **66**(1): p. 1-31.

17. Malladi, R., J.A. Sethian, and B.C. Vemuri, *Shape modeling with front propagation: A level set approach.* Pattern Analysis and Machine Intelligence, IEEE Transactions on, 1995. **17**(2): p. 158-175.

18. Suri, J.S., et al., *Shape recovery algorithms using level sets in 2-D/3-D medical imagery: A state-of-the-art review.* Information Technology in Biomedicine, IEEE Transactions on, 2002. **6**(1): p. 8-28.

19. Caselles, V., R. Kimmel, and G. Sapiro, *Geodesic active contours.* INTERNATIONAL JOURNAL OF COMPUTER VISION, 1997. **22**(1): p. 61-79.

20. Bernard, O., et al., *Variational B-spline level-set: a linear filtering approach for fast deformable model evolution.* Image Processing, IEEE Transactions on, 2009. **18**(6): p. 1179-1191.

21. Li, C., et al., *Minimization of region-scalable fitting energy for image segmentation.* Image Processing, IEEE Transactions on, 2008. **17**(10): p. 1940-1949.

22. Lankton, S. and A. Tannenbaum, *Localizing region-based active contours.* Image Processing, IEEE Transactions on, 2008. **17**(11): p. 2029-2039.

23. Shi, Y. and W.C. Karl, *A real-time algorithm for the approximation of level-set-based curve evolution.* Image Processing, IEEE Transactions on, 2008. **17**(5): p. 645-656.

24. Dice, L.R., *Measures of the amount of ecologic association between species.* Ecology, 1945. **26**(3): p. 297-302.

25. Dietenbeck, T., et al. *CREASEG: a free software for the evaluation of image segmentation algorithms based on level-set.* in *Image Processing (ICIP), 17th IEEE Int. Conf. on* 2010. Hong Kong: IEEE.

26. Rastgarpour, M., S. Alipour, and J. Shanbehzadeh, *Improved Fast Two Cycle by using KFCM Clustering for Image Segmentation.* Lecture Notes in Engineering and Computer Science: Proceedings of The International MultiConference of Engineers and Computer Scientists 2012, IMECS 2012, 2012. **1**: p. 678-682.

27. Muller, K.R., et al., *An introduction to kernel-based learning algorithms.* Neural Networks, IEEE Transactions on, 2001. **12**(2): p. 181-201.

28. Merriman, B., J.K. Bence, and S. Osher, *Diffusion generated motion by mean curvature.* 1992: Dept. of Mathematics, University of California, Los Angeles.

29. Merriman, B., J.K. Bence, and S.J. Osher, *Motion of multiple junctions: A level set approach.* Journal of computational physics, 1994. **112**(2): p. 334-363.

30. Ray, N. and S.T. Acton. *Image segmentation by curve evolution with clustering.* in *Proc. of 34th Asilomar Conf.on Sig., Sys., and Computers. Pacific.* 2000. Grove, CA, USA: IEEE.

31. *Insight Segmentation and Registration Toolkit (ITK).* Available from: http://www.itk.org/.

32. Staal, J., et al., *Ridge-based vessel segmentation in color images of the retina.* Medical Imaging, IEEE Transactions on, 2004. **23**(4): p. 501-509.

33. *Internet Brain Segmentation Repository (IBSR).* Available from: http://www.cma.mgh.harvard.edu/ibsr/.

FINGERPRINT IMAGE DEPURATION BY MULTI-STAGE COMPUTATIONAL METHOD

IWASOKUN GABRIEL BABATUNDE

Department of Computer Science,
Federal University of Technology,
PMB 704, Akure, Ondo State, Nigeria

AKINYOKUN OLUWOLE CHARLES

Department of Computer Science,
Federal University of Technology,
PMB 704, Akure, Ondo State, Nigeria

ALESE BONIFACE KAYODE

Department of Computer Science,
Federal University of Technology,
PMB 704, Akure, Ondo State, Nigeria

OLABODE OLATUBOSUN

Department of Computer Science,
Federal University of Technology,
PMB 704, Akure, Ondo State, Nigeria

Fingerprint remains a dominant biometric used for human verification and identification. This is manifesting through continuous emergence of series of Automated Fingerprint Identification Systems (AFIS). Several yardsticks including the degree of verification or identification of fingerprint images are being used to measure the performances of these systems. The quality of the fingerprint images and the efficiency of the algorithm are in turn used for the determination of this degree. This paper presents a simplified and faster substitute to one of the existing mathematical algorithm for enhancing or raising the quality of fingerprint image to the level it will be most suitable for application in any AFIS. The new algorithm comprises of phases for fingerprint ridge segmentation, normalization, orientation estimation, frequency estimation, Gabor filtering, binarization and thinning. The implementation was characterized by Window Vista Home Basic operating system as platform and Matrix Laboratory (MatLab) as frontend engine. Experiments were conducted in three phases to test the adequacies of the different sub-models of the new algorithm. The first phase was on synthetic images while the second phase was on fingerprints obtained from some selected students and staff of The Federal University of Technology, Akure, Nigeria (FUTA) using ink and paper method. The third and final phase was on standard FVC2002 fingerprint database dataset DB4. Results obtained from the three phases show that each sub-model performs well with images with free or minimal noise levels. The necessity of each stage of the enhancement is also established.

1. Introduction

Fingerprint is presently one of the essential variables used for the enforcement of security through reliable verification and identification of individuals. Fingerprints are used as elements of security during voting, operation of bank accounts among others. They are also used for access monitor to offices, equipment rooms, control centers and other highly important centers. Figure 1 presents the result of the survey conducted by the International Biometric Group (IBG) in 2004 on comparative analysis of fingerprint with other biometrics. The result shows that a substantial margin exists between the uses of fingerprint for the enforcement of security over other biometrics such as face, hand, iris, voice, signature and middleware [1].

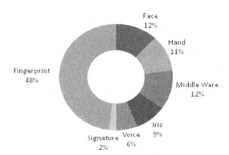

Figure 1: Comparative survey of fingerprint with other biometrics

The following reasons had been adduced for this substantial margin [1]-[4]:

a. Fingerprints have a wide variation since no two people have identical prints.
b. There is high degree of consistency in fingerprints. A person's fingerprints may change in scale but not in relative appearance, which is not the case in other biometrics.
c. Fingerprints are left each time the finger contacts a surface.

The following reasons equally account for the much larger market of personal authentication using fingerprints [5]:

a. Existence of small and inexpensive fingerprint capture devices
b. Existence of fast computing hardware
c. Existence of high recognition rate and speed equipment that meet the needs of many applications

d. The explosive growth of network and Internet transactions
e. The heightened awareness of the need for ease-of-use as an essential component of reliable security.

The components of fingerprints that are of main importance are the features points. The features exhibit uniqueness defined by type, position and orientation from fingerprint to fingerprint and they are classified into global and local features [6]-[8]. The global features are the contents of the fingerprint that could be seen with the naked eye and characterized by the attributes that capture the global spatial relationships. Notable among the global features are the ridge pattern, type, orientation, spatial frequency, curvature and position. Other global features are the core and delta areas, type lines and ridge count.

The local features also known as minutiae points, are the tiny, unique characteristics of fingerprint ridges used for positive identification. Local features contain the information that is in a local area only and are invariant with respect to global transformation. It is possible for two or more impressions of same finger to have identical global features but still differ because they have local features (minutiae points) that are different [6]. In Figure 2, ridge patterns (a) and (b) are two different impressions of the same finger (person). The same a local feature read as bifurcation in (a) appears as a ridge ending in (b). Section 2 of this paper discusses fingerprint image enhancement based on the proposed algorithm while Section 3 presents the experimental findings. The conclusion drawn is presented in Section 4.

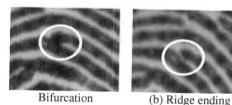

Bifurcation   (b) Ridge ending

Figure 2: Different minutiae for different impression of same finger

2. Fingerprint Image Enhancement

The first task in the use of fingerprint in an AFIS is the proper detection and extraction of features. For a sound and proper extraction of features from any fingerprint, such fingerprint is firstly enhanced. In a well enhanced fingerprint, there is clear separation between valid and spurious features. Spurious features are those foreign feature (minutiae) points that are created as noise or artifacts. This paper implements a modified version of the fingerprint enhancement

274

algorithm implemented in [9]-[10]. The conceptualized diagram of the modified algorithm is shown in Figure 3.

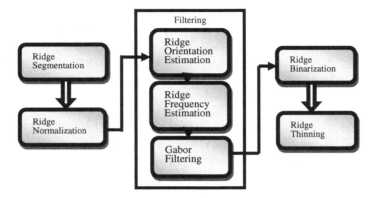

Figure 3: Conceptual diagram of fingerprint image enhancement

The algorithm provides an enhancement method that is in the following phases [5]:

2.1. *Image Segmentation*

The foreground and the background are the two regions of any fingerprint image. The foreground regions, also known as the regions of interest (RoI), contain the ridges and valleys. They are the region of interest because they encompass the feature points. The backgrounds are outside the foreground regions and they contain the noises introduced into the image during enrolment are mostly found. During segmentation, the foreground regions are separated from the background regions so the burden associated with the subsequent stages of image enhancement is lessen by ensuring that focus is only on the foreground regions.

The background regions possess very low grey-level variance values while the foreground regions posses very high grey-level variance values. Block processing approach is used in place of the previous pixel approach [9]-[10] to obtain the grey-level variance values. In block processing approach, the image is firstly divided into blocks of size W x W and the variance, V(k) for all the pixels in the kth block is calculated from:

$$V(k) = \frac{1}{W^2} \sum_{i=1}^{W} \sum_{j=1}^{W} (I(i,j) - M(k))^2 \qquad (1)$$

I(i,j) is the grey-level value for pixel i,j in block k and M(k) is the average grey-level value for block k.

2.2. *Image Normalization*

Normalization of the fingerprint image ridge structure is used to standardize the grey-level values that constitute the image. By normalization, the grey-level values are confined to a range appropriate for improved image contrast and brightness. To accomplish this, the image is divided into blocks of size S x S. A comparison is then made between the grey-level values for each pixel in the segmented image with the average grey-level value of the host block. Assumed mean of M_0 and assumed variance of V_0 are used for the comparison. For a pixel I(i,j) belonging to a block of average grey-level value of M, the normalized grey-level value N(i,j) is computed from:

$$
N(i,j) = \begin{cases} M_0 + \sqrt{\dfrac{V_0(I(i,j) - M)^2}{V}} & if \ I(i,j) > M \\ M_0 - \sqrt{\dfrac{V_0(I(i,j) - M)^2}{V}} & otherwise \end{cases} \tag{2}
$$

2.3. *Image Filtering*

Filtering of the normalized fingerprint image is done for preserving the true ridge and valley structures and for the removal of noises from the image. Fingerprint image filtering is in the following phases:

Orientation Estimation

In every fingerprint image, the ridges form patterns that flow in different directions. The direction of the flow of ridges at points A and B over range of pixels indicated by arrows shown in Fig. 4 is the ridge orientation α and β respectively.

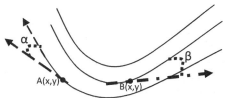

Figure 4: The orientation of ridge pixels in a fingerprint

The ridge orientations in a fingerprint image are derived using the following steps:

a. Firstly, blocks of size $S \times S$ are formed in the normalized image.
b. For each pixel, (p,q) in each block, the gradients $\partial_x(p,q)$ and $\partial_y(p,q)$ representing the gradient magnitudes in the x and y directions, respectively are computed. $\partial_x(p,q)$ is computed using the horizontal Sobel operator while $\partial_y(p,q)$ is computed using the vertical Sobel operator.

$$\begin{bmatrix} 1 & 0 & -1 \\ 2 & 0 & -2 \\ 1 & 0 & -1 \end{bmatrix} \qquad \begin{bmatrix} 1 & 2 & 1 \\ 0 & 0 & 0 \\ -1 & -2 & -1 \end{bmatrix}$$

Horizontal Sobel Operator   Vertical Sobel Operator

c. The local orientation for each pixel in the image was computed using its $S \times S$ neighborhood in [9]-[10]. This was slightly modified in this research by dividing the image into $S \times S$ blocks and the local orientation for each block centered at pixel $I(i,j)$ is then computed from:

$$V_x(i,j) = \sum_{p=i-\frac{S}{2}}^{i+\frac{S}{2}} \sum_{q=j-\frac{S}{2}}^{j+\frac{S}{2}} 2\partial_x(p,q)\partial_y(p,q) .. \tag{3}$$

$$V_y(i,j) = \sum_{p=i-\frac{S}{2}}^{i+\frac{S}{2}} \sum_{q=j-\frac{S}{2}}^{j+\frac{S}{2}} \partial_x^2(p,) - \partial_y^2(p,q) \tag{4}$$

$$\theta(i,j) = \frac{1}{2}\tan^{-1}\frac{V_y(i,j)}{V_x(i,j)} \tag{5}$$

$\Theta(i,j)$ is the Least Mean Square (LSM) estimate of the local orientation at the block centered at pixel (i,j).

d. The orientation image is then converted into a continuous vector field as follows:

$$\varphi_x(i,j) = \cos(2\theta(i,j)), \tag{6}$$

$$\varphi_y(i,j) = \sin(2\theta(i,j)), \tag{7}$$

φ_x and φ_y are the x and y components of the vector field, respectively.

e. Gaussian smoothing is then performed on the vector field as follows:

$$\varphi'_x(i,j) = \sum_{p=-\frac{S_\varphi}{2}}^{\frac{S_\varphi}{2}} \sum_{q=-\frac{S_\varphi}{2}}^{\frac{S_\varphi}{2}} G(p,q)\varphi_x\,(i-ps,j-qs) \tag{8}$$

$$\varphi'_y(i,j) = \sum_{p=-\frac{S_\varphi}{2}}^{\frac{S_\varphi}{2}} \sum_{q=-\frac{S_\varphi}{2}}^{\frac{S_\varphi}{2}} G(p,q)\varphi_y\,(i-ps,j-qs) \tag{9}$$

G is a Gaussian low-pass filter of size $S_\varphi \times S_\varphi$.

f. The orientation field O of the block centered at pixel (i,j) is finally smoothed using the equation:

$$O(i,j) = \frac{1}{2}\tan^{-1}\frac{\varphi'_y(i,j)}{\varphi'_x(i,j)} \tag{10}$$

Ridge Frequency Estimation

In any fingerprint image, there is a local frequency of the ridges that collectively form the ridge frequency image. The ridge frequency is obtained from the extraction of the ridge map. The extraction of the ridge map from any fingerprint image is in the following steps:

a. Compute the *consistency level* of the orientation field in the local neighborhood of a pixel (p,q) with the following formula:

$$C_o(p,q) = \frac{1}{n^2}\sqrt{\sum_{(i,j)\,\in\,W}|\theta(i,j)-\theta(p,q)|^2} \tag{11}$$

$$|\theta(i,j)-\theta(p,q)| = \begin{cases} d & if\ d < 180 \\ d-180 & otherwise \end{cases} \tag{12}$$

$$d = (\theta(i,j)-\theta(p,q)+360)\ mod\ 360 \tag{13}$$

W represents the n x n local neighborhood around (p,q), θ(i,j) and θ(p,q) are local ridge orientations at pixels (i,j) and (p,q) respectively.

b. If the *consistency level* is below a certain threshold F_c, then the local orientations in this region are re-estimated at a lower image resolution level until the consistency is above F_c. After obtaining the orientation field, the following two adaptive filters are then applied to the filtered image:

$$
h_t(p,q,i,j) = \begin{cases} \dfrac{-1}{\sqrt{2\pi\delta}} e^{\frac{-1}{\delta^2}}, & if\ i = l(j) - d, \quad j \in \rho \\ \dfrac{1}{\sqrt{2\pi\delta}} e^{\frac{-1}{\delta^2}}, & if\ i = l(j), \quad j \in \rho \\ 0, & otherwise, \end{cases} \tag{14}
$$

$$
h_b(p,q,i,j) = \begin{cases} \dfrac{-1}{\sqrt{2\pi\delta}} e^{\frac{-1}{\delta^2}}, & if\ i = l(j) + d, \quad j \in \rho \\ \dfrac{1}{\sqrt{2\pi\delta}} e^{\frac{-1}{\delta^2}}, & if\ i = l(j), \quad j \in \rho \\ 0, & otherwise, \end{cases} \tag{15}
$$

$$
l(j) = j\ tan\ (\theta(p,q)); \tag{16}
$$

$$
d = \frac{Y}{2\cos\ (\theta(p,q))}, \tag{17}
$$

$$
\rho = Y\left[\left|\frac{\sin(\theta(p,q))}{-2}\right|, \left|\frac{\sin(\theta(p,q))}{2}\right|\right] \tag{18}
$$

These two filters are capable of adaptively accentuating the local maximum grey level values along the normal direction of the local ridge orientation. The normalized image is first convolved with these two masks, h_t (p,q, i, j) and h_b (p,q, i, j). If both the grey level values at pixel (p,q) of the convolved images are larger than a certain threshold F_{ridge}, then pixel (p,q) is labeled as a ridge.

Gabor Filtering

Gabor filtering is used to improve the fingerprint image. It is at this stage that the removal of noise takes place. Gabor filtering is performed according to the formula:

$$
G(x,y:f,\theta) = exp\left\{\frac{1}{2}\left[\frac{a^2}{\delta_x^2} + \frac{b^2}{\delta_y^2}\right]\right\}\cos(2\pi fa) \tag{19}
$$

f is the frequency of the cosine wave along the direction θ from the x-axis, and δ_x and δ_y are the space constants along x and y axes respectively. a= $x\sin\theta$ + $y\cos\theta$ and b= $x\cos\theta$ + $y\sin\theta$.

The values of the space constants δ_x and δ_y for the Gabor filters were empirically determined as each is set to about half the average inter-ridge distance in their respective direction. δ_x and δ_y were obtained from $\delta_x = k_x F$ and $\delta_y = k_y F$ respectively. F is the ridge frequency estimate of the original image, and k_x *and* k_y are constant variables. The value of δ_x determines the degree of contrast enhancement between ridges and valleys while the value of δ_y determines the amount of smoothing applied to the ridges along the local orientation.

2.4. *Image Binarization/Thinning*

The filtered image is binarized and thinned for best performance. The image binarization technique proposed in [11] is employed with an assumed threshold (T) that minimizes overlap and maintains a clear separation among clusters. To determine the actual value of T, the following operations were performed on set of presumed threshold values:

a. The pixels are separated into two clusters according to the threshold.
b. The mean of each cluster are determined.
c. The difference between the means is squared.
d. The product of the number of pixels in one cluster and the number in the other is determined.

The success of these operations depends on the difference between the respective mean of the clusters. The optimal threshold is the one that maximizes the between-class variance or, conversely, the one that minimizes the within-class variance. The *within-class* variance of each of the cluster is calculated as the weighted sum of the variances from:

$$\sigma^2_{within}(T) = n_B(T)\sigma^2_B(T) + n_O(T)\sigma^2_O(T) \qquad (20)$$

$$n_B(T) = \sum_{i=0}^{T-1} p(i) \qquad (21)$$

$$n_O(T) = \sum_{i=T}^{N-1} p(i) \qquad (22)$$

$\sigma^2_B(T) = the\ variance\ of\ the\ pixels\ in\ the\ background\ (below)threshold$

$\sigma^2_O(T) = the\ variance\ of\ the\ pixels\ in\ the\ foreground\ (above)threshold$

p(i) is the pixel value at location i, N is the intensity level and $[0, N-1]$ is the range of intensity levels. The between-class variance, which is the difference between the within-class variance and the total variance of the combined distribution, is then obtained from:

$$\sigma^2_{between}(T) = \sigma^2 - \sigma^2_{within}(T) \tag{23}$$

$$= n_B(T)[A] + n_O(T)[B \tag{24}$$

$$A = (\mu_B(T) - \mu)^2 \tag{25}$$

$$B = (\mu_O(T) - \mu)^2 \tag{26}$$

where σ^2 is the combined variance, $\mu_B(T)$ is the combine mean for cluster T in the background threshold, $\mu_O(T)$ is the combine mean for cluster T in the foreground threshold and μ is the combined mean for the two thresholds. The between-class variance is simply the weighted variance of the cluster means themselves around the overall mean. Substituting $\mu = n_B(T)\mu_B(T) + n_O(T)\mu_O(T)$ into (24), the result is:

$$\sigma^2_{between}(T) = n_B(T)n_O(T)[\mu_B(T) - \mu_O(T)]^2 \tag{27}$$

Using the following simple recurrence relations, the between-class variance is successfully updated by manipulating each threshold T using a constant value p as follows:

$$n_B(T + 1) = n_B(T) + p \tag{28}$$

$$n_O(T + 1) = n_O(T) - p \tag{29}$$

$$\mu_B(T + 1) = \frac{\mu_B(T)n_B(T) + pT}{n_B(T + 1)} \tag{30}$$

$$\mu_O(T + 1) = \frac{\mu_O(T)n_O(T) - pT}{n_O(T + 1)} \tag{31}$$

3. Experimental Results

The proposed algorithm was implemented with MATLAB Version 7.2 on the Windows Vista Home Basic operating system using Pentium 4 – 1.87 GHz processor with 1024MB of RAM. For performance evaluation, three sets of experiments were conducted. The first set of experiments used synthetic images generated by *circsine* function [12] to verify the performance of the orientation estimation, ridge frequency estimation and Gabor filtering algorithms. The MATLAB *imnoise* function was also used to generate noise and artifacts on the synthetic images. The basic needs of these two functions are discussedd in [5]. The second set of experiments used real fingerprints images obtained from selected persons in FUTA while the third and final set of experiments was on standard FVC2002 fingerprint database dataset DB4 available on (www.bias.csr.unibo.it/fvc2002/ download.asp). The performance of the image ridge orientation algorithms on zero, medium and high level salt and pepper noise synthetic image shown in Figure 5(a), 5(b) and 5(c) are presented in Figure 5(d), 5(e) and 5(f) respectively. The synthetic images are of uniform size of 210 x 210 and wavelength 8. Figure 5(a), 5(b) and 5(c) were derived using the *imnoise* function at noise level of 0, 0.20 and 0.30 respectively. The ridge orientation estimates shown in Figure 5(d), 5(e) and 5(f) reveal the dependent of the performance of the algorithm on the image noise level. When the noise level is reasonable, the algorithm does well as shown in Figure 5(d) and 5(e). However, it produces misleading result as shown in Figure 5(f) when the noise level is beyond the threshold value of 0.29.

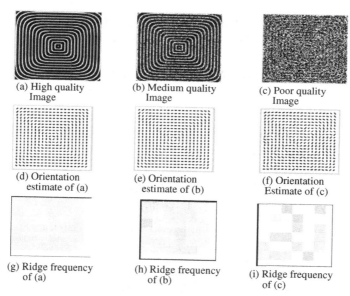

(a) High quality Image

(b) Medium quality Image

(c) Poor quality Image

(d) Orientation estimate of (a)

(e) Orientation estimate of (b)

(f) Orientation Estimate of (c)

(g) Ridge frequency of (a)

(h) Ridge frequency of (b)

(i) Ridge frequency of (c)

Figure 5: Orientation and ridge frequency estimates for synthetic images of different qualities

Figure 5(g), 5(h) and 5(i) present the results of the ridge frequency estimation experiment on the synthetic image shown in Figure 5(a), 5(b) and 5(c) respectively.

Figure 5(g) and 5(h) show uniformity between the majority of the estimated frequency values for each 32 x 32 block in the image with zero or medium noise level. Hence, it is deduced that the used algorithm is accurate for both well-defined and medium noise level images. However, visual inspection of the estimate presented in Figure 5(i) for 0.3 salt and pepper noise level image reveals that there exist a large number of image blocks with non-uniformity in wavelength estimate values. This is why it shows different pattern from the experimental results shown in Figure 5(g) and 5(h). The ridge frequency estimation algorithm diminishes in performance as the image noise density increases beyond the threshold. The performance of the Gabor filtering algorithm on a zero, medium and high quality synthetic image of size 410 x 410 and wavelength 15 is presented in Figure 6(d), 6(e) and 6(f) respectively. Parameter values of k_x =0.45 and k_y =0.45 were used to obtain these results.These results reveal that with zero or medium noise level, the filter effectively filtered the image and enhanced it to a level comparable with the original image. This is partly attributed to the accurate estimation of the ridge orientation and the ridge frequency for zero or medium noise level images. However, the experimental result presented in Figure 6(f) reveals that when the filter is applied to images of high salt and pepper noise level, the filter is unable to remove the noise effectively as it produced a significant amount of spurious features. This is attributed to inaccurate estimation of the ridge orientation and the ridge frequency arising from high noise level.

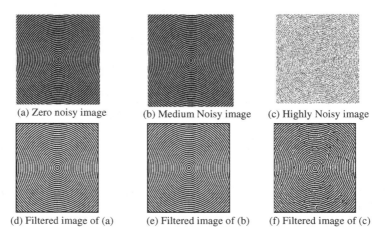

(a) Zero noisy image	(b) Medium Noisy image	(c) Highly Noisy image
(d) Filtered image of (a)	(e) Filtered image of (b)	(f) Filtered image of (c)

Figure 6: Results of applying a Gabor filter on synthetic images of different noise levels.

The best results were obtained for image segmentation using variance threshold of 100 on the second stage experiments. This threshold value provided best segmentation results in terms of differentiating between the foreground and the background regions. Figure 7(b) and 7(f) are the results of the segmentation experiments on the images shown in Figure 7(a) and 7(e) respectively.

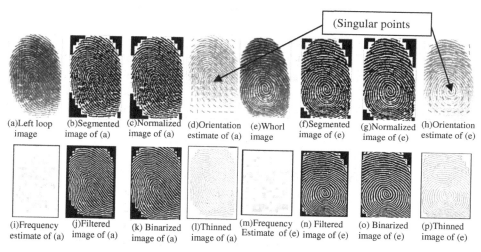

Figure 7: Results of enhancement on images

On these two results, the foreground regions are the regions containing the ridges and valleys while the background regions are the dark and outside regions. The result of the normalization experiments are presented in Figure 7(c) and 7(g) respectively. The ridges in the normalized images were normalized to a desired mean of zero and a variance of one. During normalization, the positions of the values are evenly shifted along the horizontal axis thereby making the structure of the ridges and valleys to become well and suitably positioned. The histogram plots of the original images shown in Figure 7(a) and 7(e) are presented in Figure 8(a) and 8(c) respectively. The histogram plots show that all the intensity values of the ridges in the images show irregular frequency values and also fall within the right hand side of the 0–250 scale, with no pixels in the left hand side. This leads to images with a very low contrast. The histogram plots of the normalized images of Figure 7(c) and 7(g) are presented in Figure 8(b) and 8(d) respectively.

Visual inspection of these plots reveals that the range of intensity values for the ridges has been adjusted between 0-1 scale such that there is a more evenly and balanced distribution between the dark and light pixels and that the ridge frequencies fall within close values. The plots also show that the normalization process distributes evenly the shape of the original image along the x-axis thereby making the ridges and valleys to be well and suitably positioned. This leads to images with a very high contrast as shown in Figure 7(c) and 7(g). The orientation field for the images was obtained around their singular points since they are prominent features used in any AFIS for fingerprint classification and matching.

The orientation estimates for the images shown in Figure 7(a) and 7(e) are presented in Figure 7(d) and 7(h) respectively. At the singular points, the orientation field is discontinuous and unlike the normal ridge flow pattern, the ridge orientation varies significantly. From these results, it is observed that there exists no deviation between the actual fingerprint ridge orientation and the estimated orientation of the vectors. In both cases, the algorithm produces accurate estimate of the orientation vectors such that they flow smoothly and consistently with the direction of the ridge structures in the images. In the superimposed version of images in Figure 7(d) and 7(h), the contrast of the original image is lowered in each case so as to improve the visibility of the orientation vectors. Visual inspection of the results for the ridge frequency estimation experiments shown in Figure 7(i) and 7(m), reveal that the ridge frequency estimate differs for the two images. This difference is attributed to the fact that the two fingerprints do not exhibit the same average ridge frequency characteristics because of the disparity in contrast levels as shown in the histogram plots. The intensities of frequency also differ for blocks or regions within same image as some blocks exhibit high contrast while others exhibit low contrast. Based on this, the synthetic images are more appropriate for the evaluation of the accuracy of the ridge frequency estimation algorithm.

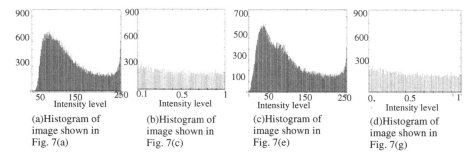

(a)Histogram of image shown in Fig. 7(a)

(b)Histogram of image shown in Fig. 7(c)

(c)Histogram of image shown in Fig. 7(e)

(d)Histogram of image shown in Fig. 7(g)

Figure 8: The histogram of original and normalized images

Figure 7(j) and 7(n) reveal how the filtering experiment was used to effectively remove noise and artifacts from the images. These results were obtained using parameter values of $k_x = 0.45$ and $k_y = 0.45$. With these values, the degree of contrast enhancement between ridges and valleys is improved and smoothing is applied to the ridges along the local orientation.

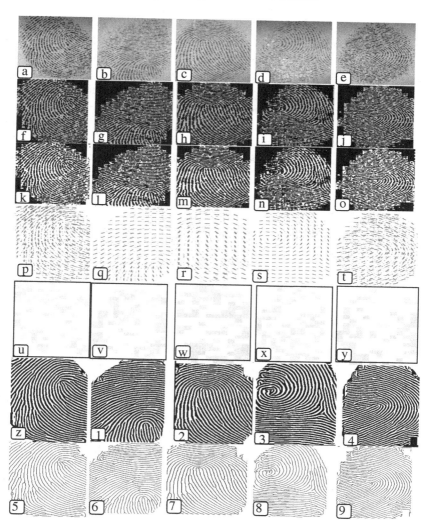

Figure 9: Results for standard fingerprint images

The results of the binarization experiments are presented in Figure 7(k) and 7(o). The binarization experiment is used to separate the ridges (black pixels) from the valleys (white pixels). The results of the thinning experiment are presented in Figure 7(l) and Figure 7(p). The MATLAB's *bwmorph* operation using the 'thin' option was used to generate the thinned images. These results show that the ridge thickness in each of the images has been reduced to its smallest form or skeleton (one pixel wide). It is also shown that the connectivity of the ridge structures is well preserved. Parts of the results of the experiments conducted at the third phase using the standard FVC2004 fingerprint database dataset DB4 are presented in Figure 9. The essence of this set of experiments is to ascertain the performance level of the different stages of the algorithm under standard fingerprint images. The results show that each stage of the algorithm does well even in image regions showing significant degradations with substantially accurate segmentation (f-j), normalization (k-o), ridge orientation (p-t), ridge frequency (u-y), Gabor filtering / binarization (z-4) and thinning results (5-9).

4. Conclusion

This report discussed the implementation of a modified version of a fingerprint enhancement algorithm proposed in [9]-[10]. Some stages of the algorithm were slightly modified for improved performance. For instance, block processing approach was introduced into the orientation estimation algorithm in place of the pixel processing approach. While the pixel processing approach subjects each pixel in the image to enhancement, the block processing approach firstly divides the image into S x S blocks before subjecting each block to enhancement. Some values were also varied from those used in [9]-[10]. For instance, parameter values of $k_x = 0.45$ and $k_y = 0.45$ were found to perform well in the image filtering experiments as against $k_x = 0.5$ and $k_y = 0.5$ used in [9]-[10].

The results of the 3-phase experiments conducted for image segmentation, normalization, ridge orientation estimation, ridge frequency estimation, Gabor filtering, binarization and thinning on synthetic and local and standard fingerprint images reveal that with free or minimal noise level, the algorithms perform well. The results also show that each stage of the enhancement process is important for obtaining a finally enhanced image that is acceptable and presentable to the next stage of fingerprint processing which is features extraction. The results obtained from the final stage of thinning show that the connectivity of the image ridge structure has been preserved at each stage.

References

1. C. Roberts, 'Biometrics'
   http://www.ccip.govt.nz/newsroom/informoationnotes/2005/biometrics.pdf,
   Accessed: July, 2009
2. C. Michael and E. Imwinkelried, 'Defence   practice tips, a cautionary note
   about fingerprint analysis and reliance on digital technology', Public
   Defense Backup Centre Report (2006)
3. M. J. Palmiotto, 'Criminal Investigation'. Chicago: Nelson Hall, 1994
4. D. Salter, 'Fingerprint – An Emerging Technology', Engineering
   Technology, New Mexico State University (2006)
5. G. B. Iwasokun, O. C. Akinyokun, B. K. Alese and O. Olabode, A Multi-
   Level Model for Fingerprint Image Enhancement, Lecture Notes in
   Engineering and Computer Science: Proceedings of The International
   MultiConference of Engineers and Computer Scientists 2012, IMECS 2012,
   14-16 March, 2012, Hong Kong, pp 730-735
6. J. Tsai-Yang, and V. Govindaraju, A minutia-based partial fingerprint
   recognition system, Center for Unified Biometrics and Sensors, University
   at Buffalo, State University of New York, Amherst, NY USA 14228 (2006)
7. O. C. Akinyokun, C. O. Angaye and G. B. Iwasokun, 'A Framework for
   Fingerprint Forensic'; Proceeding of the First International Conference on
   Software Engineering and Intelligent System, organized and sponsored by
   School of Science and Technology, Covenant University, Ota, Nigeria
   (2010), pages 183-200.
8. O. C. Akinyokun and E. O. Adegbeyeni, 'Scientific Evaluation of the
   Process of Scanning and Forensic Analysis of Fingerprints on Ballot
   Papers', Proceedings of Academy of Legal, Ethical and Regulatory Issues,
   Vol. 13, Numbers 1, New Orleans(2009).
9. L. Hong, Y. Wan and A. Jain, 'Fingerprint image enhancement: Algorithm
   and performance evaluation'; Pattern Recognition and Image Processing
   Laboratory, Department of Computer Science, Michigan State University,
   (2006), pp1-30
10. T. Raymond, Fingerprint Image Enhancement and Minutiae Extraction,
    PhD Thesis Submitted to School of Computer Science and Software
    Engineering, University of Western Australia (2003).
11. X. Liang, 'Image Binarization using Otsu Method', Proceedings of NLPR-
    PAL Group CASIA Conference (2009), pp345-349
12. P Kovesi, 'MATLAB functions for computer vision and image analysis',
    School of Computer Science and Software Engineering, University of
    Western Australia, http://www.cs.uwa.edu.au/~pk/Research/MatlabFns/
    Index.html, Accessed: 20 February 2010.

HUMAN BIO FUNCTIONS AS FPGAs CHIP DESIGN
- AN INSULIN PERSPECTIVE

AMMAR EL HASSAN

College of Computer Engineering & Science
Prince Muhammad bin Fahd University
AL-Khobar, Saudi Arabia
aelhassan@pmu.edu.sa

LOAY ALZUBAIDI

College of Computer Engineering & Science
Prince Muhammad bin Fahd University
AL-Khobar, Saudi Arabia
lalzubaidi@pmu.edu.sa

JAAFAR AL GHAZO

College of Computer Engineering & Science
Prince Muhammad bin Fahd University
AL-Khobar, Saudi Arabia
jghazo@pmu.edu.sa

Abstract.   In this article, which is based on a paper by the same authors [16], the modeling and synthesis of human Insulin Hormone Secretion Mechanism is accomplished using VHDL and FPGAs technologies.   A mathematical model is developed and analyzed using Matlab and Least-Square fitting algorithm.   C++ is used to model the behavior of Insulin secretion in humans and converted to VHDL. Results are verified then the mechanism is realized on a Xilinx FPGAs chip. This chip is then tested with simulated input and its behavior is deemed consistent with the mathematical model. The chip is therefore an identical replica of the Human Insulin Secretion Mechanism.

1. Introduction

When evaluating new drugs or treatment techniques for common diseases in animals or humans, there are numerous benefits inherent in the availability and use of   artificial or synthetic systems that mimic biological mechanisms and functions. These benefits range from costs to ethics and make it much more feasible for new drugs and innovative disease treatment techniques to be assessed risk-free [1] and [3].

Field Programmable Gate Arrays (FPGAs) are a new technology that provides users programmability in the field. They contain arrays of Configurable Logic Blocks (CLBs) that can be programmed to realize different designs. FPGA families differ by their chip-level architecture and by the granularity of the function unit and intra- and inter-chip wiring organizations. Because of the short turnaround time and low cost, there is increased interest in system prototyping using FPGAs [4]. Along with the right programming environment such as VHDL, FPGAs are a very powerful tool for the design and testing of system prototypes.

VHDL is used to model the secretion mechanism of the Insulin hormone in humans; including the effects of Glucose and Glucagon levels. Xilinx tools are utilized to realize the model onto a FPGA chip. VHDL tools can model the system and realize it onto Integrated Circuit. The model can be simulated and displayed on computer screens to examine the results. Once the results satisfy the users, CAD tools are used to implement the design and produce a bit file to program the FPGAs chip. Once programmed, the FPGAs chip mimics the system.

In this article, we introduce and demonstrate a technique for the modeling, design and simulation of the behavior of the Insulin hormone in humans. We design a compact integrated circuit chip using various Computer Aided Design (CAD) tools including hardware-descriptive language techniques including Verilog /VHDL); which are hardware description languages to design digital logic using FPGAs. FPGAs provide optimal device utilization through conservation of board space and system power.

The applications for this technology are wide and varying; by offering a platform to simulate the behavior of human Insulin levels in response to varying conditions and drugs, it is hoped that a lot more can be learned about Insulin and new drugs can be produced to help regulate this hormone in humans with diabetes and other medical conditions in an ethical way that preserves the rights and dignity of human subjects.

The chip introduced in this paper can be used in 2 distinct ways:

o   By connecting the chip to a human subject, it can receive electrical signals from sensors attached to various regions of the body such as brain, heart, or muscles.   These analog signals are converted using an Analog-to-Digital (A/D) converter. The effect of various drugs on the body is thus detected by the sensors and the output signals of the sensors are transmitted to the input pins of the chip. The output pins of the chip display the output of the mechanism for the given input signals. The output signals can be re-converted to analog using Digital-to-Analog (D/A) converters. This is illustrated in Fig.1.

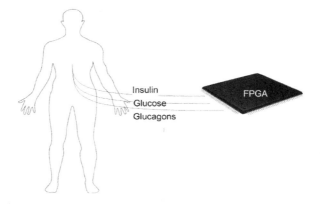

Fig. 1. The FPGAs chip with a human subject

o The other option excludes the human subject from the setup and relies instead on signal/function generators to simulate the human body output sensors, thus allowing the realized mechanism on the chip to be tested extensively and ethically in a lab environment with no human contact. This is illustrated in Fig. 2.

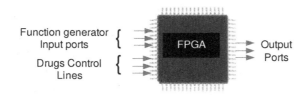

Fig. 2. The FPGAs chip with no human subject

Blood Glucose and Insulin

For their energy needs, the cells in our bodies rely on the breakdown of Glucose which is absorbed from food via the intestines and consequently circulated via the blood stream.

Because food intake, and hence Glucose levels in the blood stream are not constant, it is imperative that this is regulated, hence our bodies store (synthesize) [11, 13] excess Glucose in the muscles and liver as Glycogen, sometimes referred to as animal starch [10]. This can later be called upon (re-synthesized) when there is a shortage of Glucose that is not obtainable from food. The Glucose-to-Glycogen transformation is essential to prevent Glucose overdose in the cell, for example after meals, the Glycogen-to-Glucose transformation is essential to prevent Glucose shortages in the cell, for example overnight. To achieve the target of a constant blood-Glucose level, our bodies rely on two hormones produced in the Pancreas that have opposite effects: Insulin and Glucagon.

Insulin, which is a protein hormone that carries 51 amino acids and originates in the Pancreas, is required by almost all of the body's cells to prevent excess Glucose. Insulin plays a particularly significant role in the control of Glucose levels in liver, and muscle cells. For these cells, Insulin is the catalyst in the Glucose-to-Glycogen transformation process (synthesis).

There are other functions that are triggered or regulated by Insulin including the formation of fats from fatty acids within fat cells, the creation of proteins from amino acids and the prevention of the liver and kidneys from making Glucose from intermediate compounds of metabolic pathways (*gluconeogenesis*). These are beyond the scope of this work and shall be disregarded at this; this is in addition to the assumptions in section 2 below.

To sum up, the post-meal Glucose level spikes (*hyperglycemia*) are effectively countered with corresponding Insulin increments [12] while extreme drops in Glucose levels (*hypoglycemia*) due to fasting, starvation, sleep, low-carbohydrate diet or intense exercise are countered with Glucagon, which is

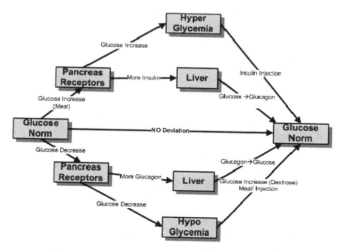

Fig. 3. Insulin & Glucagon as Glucose Controls

another hormone secreted by the Pancreas, that raises blood Glucose levels. Hence, its effect is opposite to that of Insulin [13].

2. Background Work & Assumptions

This work is inspired by and builds upon similar work undertaken earlier by [1, 2 & 3] which utilized VHDL, FPGAs and other hardware realization tools to model, build and test an integrated chip that mimics the behavioral patterns of the Human Growth Hormone (HRH) amongst others.

In terms of assumptions, the writers understand, but will sidetrack factors including: Insulin level variations by test region (abdomen vs. extremities vs. brain vs. liver) or the effect of factors (like growth hormone or deficiency thereof) on the levels of Insulin [6]. The effect of other non-biological factors (e.g., alcoholism) are also acknowledged but not covered by this work [7]. The main factor that plays a significant role on the observed behavior pattern of Insulin levels in plasma is the effect of meals and time of day/night on this measurement [8]. The raw data used for this work shows the behavior of Insulin over a time-span of 24 hours, hence this is the time-span with which this work is concerned. For a longer time-span, further work will be required to ascertain the existence of additional patterns of behavior (which will not be covered here) or merely cyclical patterns of behavior (which will implicitly be covered). Other restrictions to the mechanism parameters here exclude the effects of fasting on Insulin release/levels [9].

These restrictions are necessary for the purpose of focusing on the salient point of the research; which is the modeling, design, implementation and simulation of an FPGAs chip that mimics the behavior of Insulin in human plasma. If this work is successful then it will provide impetus for further studies with those additional factors included.

3. Process

The steps that were undertaken in this research are illustrated in Fig. 3 and are explained in this section.

The technology of FPGAs provides a programmable interface to enable us to synthesize complex behavior models. FPGAs chips Configurable Logic Blocks (CLBs) can be tailored to represent different models [1]. The three distinguishing features of FPGAs chip: ***architecture, function-unit granularity and intra/inter-chip wiring organization*** can be fine-tuned to represent complex models in a fairly short period of time. Combined with VHDL tools, we have a powerful tool to represent and synthesize our mathematical model.

The input to Matlab is the raw data representing the secretion patterns of Insulin (in response to fluctuations in Glucose & Glucagon levels). The model in Fig. 5 shows this behavior over a 24-hour time span.

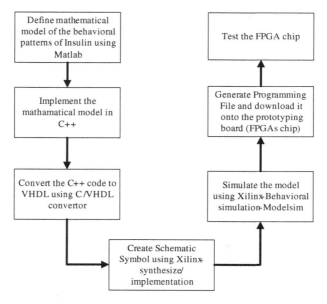

Fig. 4. Mathematical Models to FPGA Chip Model Process

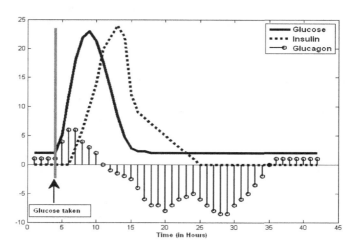

Fig. 5. Insulin, Glucose & Glucagon
production in the blood stream

After several trials with the graphics facility using Matlab, the model was fitted using the ***non-linear least-square*** [14,15] algorithm, as shown in the equations below:

$$Insulin\ (t) = a1 * sin(b1 * t + c1) + a2 * sin(b2 * t + c2) + a3 * sin(b3 * t + c3) + a4 * sin(b4 * t + c4) + a5 * sin(b5 * t + c5);$$

Eq. 1

$$Glucose\ (t) = ga1 * sin(gb1 * t + gc1) + ga2 * sin(gb2 * t + gc2) + ga3 * sin(gb3 * t + gc3) + ga4 * sin(gb4 * t + gc4) + ga5 * sin(gb5 * t + gc5) + ga6 * sin(gb6 * t + gc6) + ga7 * sin(gb7 * t + gc7) + ga8 * sin(gb8 * t + gc8);$$

Eq. 2

$$Glucagon\ (t) = gga1 * sin(ggb1 * t + ggc1) + gga2 * sin(ggb2 * t + ggc2) + gga3 * sin(ggb3 * t + ggc3) + gga4 * sin(ggb4 * t + ggc4) + gga5 * sin(ggb5 * t + ggc5) + gga6 * sin(ggb6 * t + ggc6) + gga7 * sin(ggb7 * t + ggc7) + gga8 * sin(ggb8 * t + ggc8);$$

Eq. 3

where:

T	time in hours
a1, a2, a3, a4, a5	fitting parameters for Insulin in mg/dl (milligram /deciliter)
b1, b2, b3, b4, b5	dimensionless fitting parameters for Insulin
c1, c2, c3, c4, c5	fitting parameters for Insulin in hours
ga1, ga2, ga3, ga4, ga5, ga6, ga7, ga8	fitting parameters for Glucose in mg/dl
gb1, gb2, gb3, gb4, gb5, gb6, gb7, gb8	dimensionless fitting parameters for Glucose
gc1, gc2, gc3, gc4, gc5, gc6, gc7, gc8	fitting parameters for Glucose in hours
gga1, gga2, gga3, gga4, gga5, gga6, gga7, gga8	fitting parameters for Glucagon in mg/dl
ggb1, ggb2, ggb3, ggb4, ggb5, ggb6, ggb7, ggb8	dimensionless fitting parameters for Glucagon
ggc1, ggc2, ggc3, ggc4, ggc5, ggc6, ggc7, ggc8	fitting parameters for Glucagon in hours

Eq's 1, 2 & 3 – Fitted Mathematical Model representing the Dynamic Behavior Patterns of Insulin, Glucose & Glucagon Respectively

A C++ program was written in order to **implement the mathematical model** of Insulin secretion including the link to Glucose and Glucagon (Eq. 1, 2, and 3), the program was executed using the values of the fitting parameters (which were generated by Matlab).

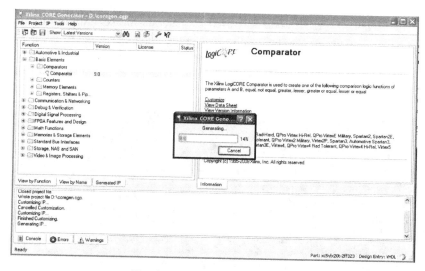

Fig. 6a. C to VHDL using Xilinx SE

ENTITY Insulin_func IS
port(

clock :	**IN**	**std_logic ;**
Glucose :	**OUT**	real ;
Insulin :	**OUT**	real;
Glucagon :	**OUT**	real;
time_hours :	**INOUT**	real;
Glucose _taken:	**INOUT**	std_logic;
done :	**INOUT**	std_logic);

END Insulin_func;

where:

clock:	input in hours
time_hours:	count the number of pulses
Glucose :	behavior of Glucose
Insulin :	behavior of Insulin
Glucagon :	behavior of Glucagon
Glucose _taken:	a trigger to start the process
done:	indicates the end of the process

Fig. 6b. VHDL Code generated by Spark 1.3

The VHDL code was then generated using **C to VHDL program** (Spark 1.3), wherein 60% of the code was generated by the convertor, the generated code was then edited for correction and further optimization, the VHDL code was subsequently compiled using the Xilinx toolset within Matlab (Fig. 6a & 6b),

The compiled VHDL code was used to generate the **schematic diagram** using Xilinx which can also be used to realize the implementation of the model into Field programmable gate array (FPGAs). The schematic diagram in Fig. 7 was generated from the VHDL code below which describes the input and output pins of the developed model.

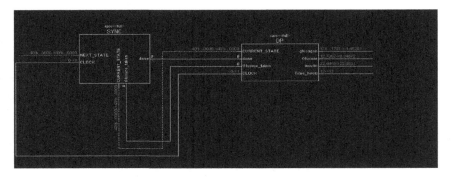

Fig. 7. Schematic Diagram of the developed integrated circuit

4. Results - Simulation of the Behavioral Model using Modelsim

The ModelSim toolkit is used for timing simulation which allows us to take the VHDL file of the previous step and use it to simulate the behavioral model. The following steps describe the simulation process:

  i. The process will be started by a trigger corresponding to the Glucose_taken input pin.
 ii. The Glucose starts to increase.
iii. This is followed by corresponding increases in Insulin and Glucagon (output pins).
 iv. Glucose reaches the MAX value (23 mg/dl) after 5 hours; Insulin reaches the MAX value (24.5 mg/dl) after 14 hours. These values correspond to the data represented in Fig. 5.

The results of the simulation are illustrated in Fig. 8 (a & b) with each pulse corresponding to a single hour. The output of this process is consistent with the simulation results and the model of Fig. 5 above as illustrated in Table 1 which shows the raw data for the model in Fig. 5.

Finally, a bit-file was generated (using Xilinx toolset) and loaded to the FPGAs chip, after testing the chip with this file, the observed results at the output pins of the chip were found to be consistent with the results of the simulation process in Fig's 8a and 8b.

Table 1 the simulation outputs of the model

Clock #	Glucose	Insulin	Glucagon
1	1.94022	0.139586	0.988018
2	2.20565	-0.242261	0.857824
3	1.53008	-0.097909	0.960493
4	2.29797	0.0908576	1.87422
5	5.83729	0.210925	3.8285
6	11.6034	0.667952	5.61706
7	17.6654	2.18048	5.77227
8	21.9154	5.30944	4.40463
9	23.1252	9.98854	2.77541
10	21.3066	15.3459	1.48028
11	17.4043	19.959	0.243453
12	12.7072	22.4435	-0.874752
13	8.34912	22.0807	-1.45498
14	5.04995	19.1528	-1.8151
15	3.05548	14.8044	-2.80759
16	2.19984	10.5044	-4.42101
17	2.05548	7.38829	-5.76012
18	2.14783	5.82138	-6.50887
19	2.16259	5.38369	-7.18259
20	2.03941	5.24438	-7.75458
21	1.89982	4.68526	-7.38991
22	1.87462	3.47557	-6.01495
23	1.9719	1.91258	-4.91322
24	2.08417	0.550801	-5.10151
25	2.10502	-0.180659	-6.13497
26	2.03126	-0.23	-7.13618
27	1.95288	0.0934501	-7.95456
28	1.95068	0.361243	-8.57145
29	2.01251	0.313617	-8.41217
30	2.05398	0.000135618	-7.22304
31	2.01693	-0.307231	-5.79735
32	1.93662	-0.350275	-4.8349
33	1.903	-0.0930237	-3.81382
34	1.9579	0.249009	-2.02803
35	2.04418	0.379735	-0.056115
36	2.06932	0.167725	0.973931
37	2.01552	-0.222227	1.06367

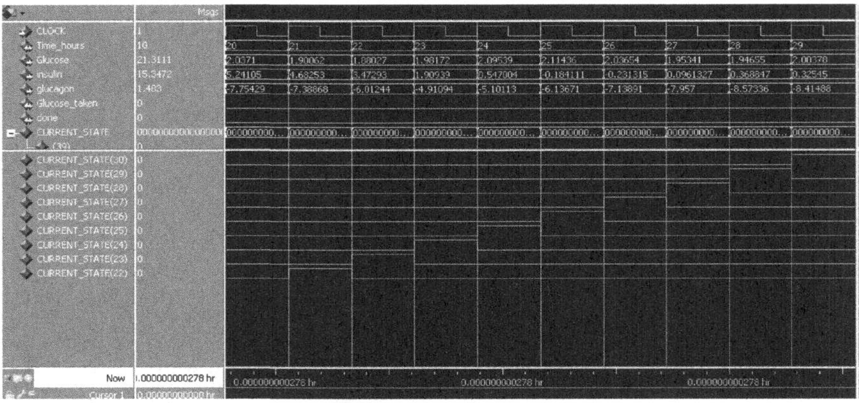

Fig. 8a. The simulation results of ModelSim – 20HRS to 29HRS

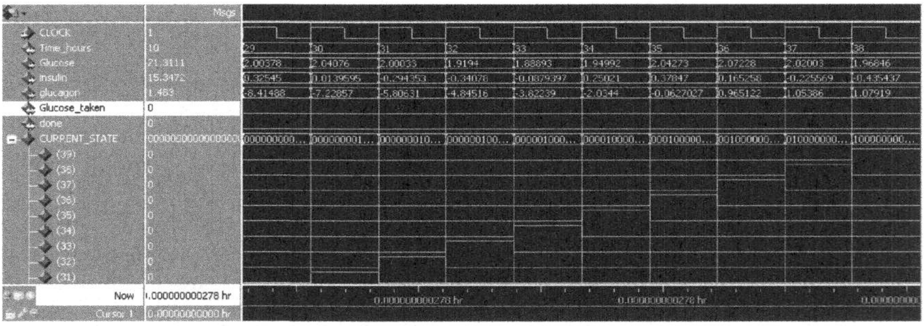

Fig. 8b. The simulation results of ModelSim – 29HRS to 38HRS

5. Conclusions & Future Work

We have successfully formulated a mathematical model that describes the Insulin secretion patterns in humans, then used this model to generate a blueprint for a microchip and generated a bit file to the prototyping board and produced some simulated hormone level figures from the chip.

There are a few restrictions with which this work started, some of which can be overcome with similar further work. Some of the areas that could be re-addressed include the mathematical model; the model used in this work represents the behavior of Insulin over a determined period of time as an observed and recorded pattern, further work is needed here to present a more interactive (dynamic) behavior model and hence, chip design which is even truer to form. Other work limitations have been listed above (section 2) and constitute good opportunities for further investigation which will be addressed in future work.

References

1. Alghazo, J., Akaaboune, A., Botros, N., "Modelling, Synthesis and Realization of Hgh Mechanism UsingVHDL and FPGAs," International Journal of Modeling and Simulation, 2005.

2. Alghazo J., Aboueida W. and Rabadi W. "Modeling and Synthesis of Thyroid Hormones Secretion Mechanism Using CAD tools" IMIBE594

3. Botros, N., Akaaboune, M., Alghazo, J., and Alhreish, M., "Hardware Realization of Biological Mechanisms Using VHDL and FPGAs," Proceedings of the Third International Conference on Modeling and Simulation of Microsystems, San Diego, CA, USA, March 27-30, pp. 233-236, 2000.

4. Botros, N., Akaaboune, A., Alghazo, J., " Modeling and Synthesis of Human Growth Hormone Secretion Mechanism Using CAD Tools," Proceedings of the 29th annual Conference of the IEEE Industrial Electronics Society, Roanoke, Virginia, USA, November 2-6, pp. 2429-2434, 2003.

5. William F. Ganong, *Review of Medical Physiology, 15th edition.* Univ of California, San Francisco: Prentice Hall International Inc, 1991.

6. Groop L., Segerlantz M. and Bramnert M. "Insulin Sensitivity in Adults with Growth Hormone Deficiency and Effect of Growth Hormone Treatment". Horm Res 2005;64(suppl 3): 45–50. On-line January 2006

7. Addolorato G., Leggio L., Hilemacher T., Kraus T., Jerlhag E., Bleich S. "Hormones and drinking behavior: New findings on ghrelin, Insulin, leptin and volume-regulating hormones. An ESBRA Symposium report". Drug and Alcohol Review (March 2009), 28, 160–165. DOI: 10.1111/j.1465-3362.2008.00023.x

8. Malherbe C., de Gasparo M. , de Hertogh R. and Hoem J. J. "Circadian variations of blood sugar and plasma Insulin levels in man". Presented at the Fourth Annual Meeting of the European Association for the Study of Diabetes, Louvain, July 23, 1968. Online in DIABETOLOGIA Volume 5, Number 6, 397-404, DOI: 10.1007/BF00427978

9. Juhl C., Grofte T., Butler P., Veldhuis J., Schmitz O. and Pørksen N. "Effects of Fasting on Physiologically Pulsatile Insulin Release in Healthy Humans". Diabetes February 2002 vol. 51 no. suppl 1 S255-S25, http://diabetes.diabetesjournals.org/content/51/suppl_1/S255.full

10. Saladin, Kenneth S. "Anatomy and Physiology". McGraw-Hill, 2007.

11. Pedersen DJ, Lessard SJ, Coffey VG, et al. (July 2008). "High rates of muscle Glycogen resynthesis after exhaustive exercise when carbohydrate is coingested with caffeine". Journal of Applied Physiology 105 (1): 7–13

12. Freudenrich C., Ph.D. "Diabetes Overview" 22 June 2001. HowStuffWorks.com. <http://health.howstuffworks.com/diseases-conditions/diabetes/diabetes.htm> 23 October 2011

13. Reece J, Campbell N. (2002) "Biology". San Francisco: Benjamin Cummings. ISBN 0-8053-6624-5.

14. Kelley C. T. "Iterative Methods for Optimization", SIAM Frontiers in Applied Mathematics, no 18, 1999

15. Strutz T. "Data Fitting and Uncertainty (A practical introduction to weighted least squares and beyond)". Vieweg+Teubner, ISBN 978-3-8348-1022-9

16. Elhassan A., Al Zubaidi L. Al Ghazo J. "Modeling and Synthesis of Human Insulin Secretion Mechanism Using CAD". Proceedings of The International MultiConference of Engineers and Computer Scientists 2012 (IMECS 2012), 14-16 March, 2012, Hong Kong, pp117-122

HAMAKER COEFFICIENT CONCEPT APPROACH AS A SURFACE THERMODYNAMIC TOOL FOR INTERPRETING THE INTERACTION MECHANISMS OF HUMAN IMMUNODEFICIENCY VIRUS AND THE LYMPHOCYTES

C.H. ACHEBE

Department of Mechanical Engineering, Nnamdi Azikiwe University, P.M.B 5025 Awka, Anambra State, Nigeria, chinobert2k@yahoo.com

S.N. OMENYI

Department of Mechanical Engineering, Nnamdi Azikiwe University, P.M.B 5025 Awka, Anambra State, Nigeria, omenyinj@hotmail.com

Sequel to the earlier works by Omenyi et al which established the role of surface thermodynamics in various biological processes from the electrostatic repulsion and van der Waals attraction mechanisms, HIV-blood interactions were modelled. This involved the use of the Hamaker coefficient approach as a thermodynamic tool in determining the interaction processes. It therefore became necessary to apply the Lifshitz derivation for van der Waals forces as an alternative to the contact angle approach which has been widely used in other biological systems. The methodology involved taking blood samples from HIV-infected and uninfected persons for absorbance measurement using Ultraviolet Visible Spectrophotometer. From the absorbance data various variables required for computations with the Lifshitz formula were derived. The Hamaker constants A_{11}, A_{22}, A_{33} and the combined Hamaker coefficients A_{132} were obtained using the values of the dielectric constant together with the Lifshitz equation. The absolute combined Hamaker coefficient, A_{132abs} for the infected blood samples gave the value of 0.2587×10^{-21} Joule. The positive sense of this value implies net positive van der Waals forces indicating an attraction between the virus and the lymphocyte. A lower value of $A_{131abs} = 0.1026 \times 10^{-21}$ Joule obtained for the uninfected blood samples is also an indicator that a zero or negative absolute combined Hamaker coefficient is attainable.

1. Introduction

1.1. *Rationale*

At the 2001 Special Session of the UN General Assembly on AIDS, 189 nations agreed that AIDS was a national and international development issue of the highest priority [1]. Between December 2005 and March 2006, UNAIDS compiled data from reports obtained from 126 countries on HIV/AIDS prevalence. In sub-Saharan Africa a mature epidemic continues to ravage beyond limits that many experts believed impossible. Also, relatively new but rapidly growing epidemics in regions such as Eastern Europe and South-East Asia that may come to rival that of sub-Saharan Africa in scope, had erupted [2].

The Highly Active Anti-retroviral Therapy (HAART) has not actually shown an easy and comprehensive solution due to the rapid mutative genetic nature of the virus [3]. The choice to approach it via the vehicle of surface thermodynamics against the conventional clinical methods is a novel one. The role of surface properties in various biological processes is now well established. In particular, interfacial tensions have been shown to play an important, if not crucial role in phenomena as diverse as the critical closing and opening of vessels in the microcirculation, cell adhesion, protein adsorption, antigen-antibody interactions, and phagocytosis [4].

1.2. *Background to Study*

The HIV is assumed to be a particle which is dispersed in a liquid (the serum) and attacks another particle (the lymphocytes). The initial actions take place on the surfaces of the cell and of the virus (assumed to be particles). This interaction which involves two surfaces coming together in the first instance can be viewed as a surface effect.

It therefore stands to reason that, if it is possible to determine the surface properties of the interacting particles, then one can predict the mechanisms of their interactions. When two particles make contact, they establish a common area of contact. Some original area of the surface of each particle has been displaced, and the work done to displace a unit area of the surface is referred to as the surface free energy. The actions therefore that take place on the surfaces are termed surface thermodynamic effects. These actions are assumed to occur slowly so that thermodynamic equilibrium is assured. This concept will be employed in this research work to characterize the HIV-blood interactions with the serum as the intervening medium.

The clinicians have analyzed the surfaces of blood cells on which the virus binds. There are receptors and co-receptors on these cells and figure 1 shows the interactions of these cells. The CCR5 is the preferred co-receptor for HIV in the human immune system. CCR5 is a seven trans-membrane protein or 7TM which means that it crosses the plasma membrane of the cell seven times. They communicate what happens outside the cell to the inside of the cell through a process called alosterism. The Chemokines bind to CCR5 which causes the CCR5 to change shape both outside and inside of the cell. The altered shape of

CCR5 changes the interactions with G-proteins inside the cell initiating a signal transduction cascade known as chemotaxis that activates the cell to go to the site of injury.

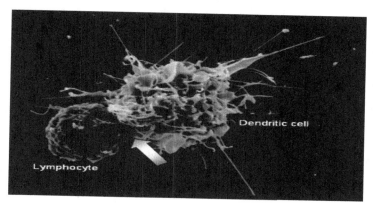

Figure 1. Interaction of a dendritic cell (right) having HIV bound to its surface (arrow) with a lymphocyte (left) [5]

The redundancy inherent in the immune system allows many Chemokines to signal for multiple coreceptors. CCR5 binds the Chemokine's RANTES, MIP-1α and MIP-1β. It is important to note that these Chemokines also bind to other receptors. Both RANTES and MIP-1α can bind to CCR1 and RANTES can also bind to CCR3. This is an example of redundancy which is common in the immune system.

1.3. Statement of Problem

HIV, being an RNA-based rapidly mutating virus, (unlike the DNA-based counterparts) lacks the ability to check for and correct genetic mutations that can occur during replication. In chronic HIV cases, about ten billion new viral species can be generated daily. This rapid genetic variation has made it rather very difficult to proffer a clinical solution to the problem [3] and the worldwide picture is one of increasing rates of infection [6]. It is against this backdrop that this study explores a novel and rare approach to the topic of HIV-blood interactions [4].

1.4. Objective of the Study

In this research work the following tasks, must be kept in view;

(i)   Determine the mechanism of interaction of HIV with white blood cells.

(ii)   Seek a thermodynamic interpretation of such interactions through van der Waals attraction mechanism.

(iii) Quantify such interactions through actual measurements.

(iv) Recommend possible approach to eliminating the HIV-blood interactions.

2. Theoretical Considerations

2.1. *Concept of Interfacial Free Energy*

The work done by a force F to move a flat plate along another surface by a distance dx is given, for a reversible process, by;

$$\delta w = Fdx \qquad (1)$$

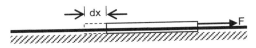

Figure 2. Schematic diagram showing application of a force on a surface

However, the force F is given by;

$$F = L\gamma \qquad (2)$$

Where L is the width of the plate and γ is the surface free energy per unit surface area (interfacial free energy)

Hence; $\delta w = L\gamma dx$

But; $dA = Ldx$ (3)

Therefore; $\delta w = \gamma dA$

This is the work required to form a new surface of area dA. For pure materials, γ is a function of T only, and the surface is considered a thermodynamic system.

2.2. *Thermodynamic Approach to Particle-Particle Interaction*

Consider the case where the virus, HIV conceived as a particle approaches the CD4 lymphocyte (also assumed to be a particle) and attaches itself on the surface of the lymphocyte dispersed in a serum, as shown in figure 3.

Figure 3. Schematic of HIV-CD4 lymphocyte adhesion process as a particle-particle interaction

The thermodynamic free energy of adhesion, ΔF^{adh} for the process shown in figure 4 can be expressed as follows [7].

$$\Delta F^{adh} = \gamma_{ps} - \gamma_{pl} - \gamma_{sl} \qquad (4)$$

Where the subscript P stands for the virus, S stands for the blood cell and L the serum. ΔF^{adh} is the free energy of adhesion integrated from infinity to the equilibrium distance, d_o. For the virus to succeed in penetrating the membrane of the blood cells, the net free energy of engulfing (of the virus by the blood cell) will be given by;

$$\Delta F_{NET} = \gamma_{ps} - \gamma_{pl} < 0 \qquad (5)$$

If ΔF_{NET} is greater than zero, the blood cell membrane will reject the virus. The drawback in the use of Eq.(5) is the difficulty in measuring the solid-liquid interfacial tension γ_{sl}. Neuman [8] in his rigorous work solved the problem by providing an empirical expression given in Eq. (6).

$$\gamma_{sl} = \frac{\left(\gamma_{sv}^{1/2} - \gamma_{lv}^{1/2}\right)^2}{1 - 0.015\left(\gamma_{sv}\gamma_{lv}\right)^{1/2}} \qquad (6)$$

The calculation of dispersion interactions from Liftshitz theory is independent of contact angle data and these interactions are not approximated by geometric means as is the case when surface tension component equations are used [9].

2.3. Lifshitz Theory

The free energy of interaction for a system consisting of two plane, semi-infinite, parallel bodies of materials 1 and 2 separated by a material 3, of thickness L is given by;

$$\Delta F_{132} = -\left[\frac{-A_{132}}{12\pi L^2}\right] \qquad (7)$$

Where, A_{132} is the Hamaker constant for the system. The Hamaker constant can be calculated through the pair-wise additivity approach as originally proposed by Hamaker [10] or by the macroscopic approach of Lifshitz [11]. The Hamaker coefficient is therefore, the macroscopic resultant of the interactions due to the polarization of the different atoms in the material [9]. The Hamaker coefficient, according to Lifshitz theory is given by:

$$A_{PLS} = \frac{3}{4}\pi\hbar\int_0^\infty \left[\frac{\varepsilon_i(i\zeta)-\varepsilon_k(i\zeta)}{\varepsilon_i(i\zeta)+\varepsilon_k(i\zeta)}\right]\left[\frac{\varepsilon_j(i\zeta)-\varepsilon_k(i\zeta)}{\varepsilon_j(i\zeta)+\varepsilon_k(i\zeta)}\right]d\zeta \qquad (8)$$

Where $\varepsilon_j(i\zeta)$ is the dielectric constant of material, j along the imaginary frequency axis and $\hbar$ is Planck's constant. Isrealachvili [12] introduced a cut-off distance parameters d_o, which represents the closest distance that two surfaces can approach. When the surfaces are at a distance d_o, they are considered to be in molecular contact. The parameter d_o, therefore eliminates the divergence in Lifshitz theory. The free energy of adhesion, using the concept of d_o, is related to Eq. (7) via:

$$\Delta F^{adh} = -\left[\frac{A_{PLS}}{12\pi d_0^2}\right] \qquad (9)$$

Hough and White [9] found that a value of 1.6×10^{-10}m gave satisfactory estimates of interfacial tensions of liquid alkanes while Omenyi et al found that a value of 1.82×10^{-10}m was satisfactory in various particle interaction processes [3]. The lifshitz-van der Waals constant A_{132} otherwise known as Hamaker coefficient could be negative. In such instance a repulsive electrostatic force will be developed which impairs contact between the interacting particles.

Table 1. Combinations of materials for which negative Lifshitz-van der Waals constant A_{132} is found [13].

System	A_{132} /eV
Si/Al$_2$O$_3$	- 0.19
Ge/Cds/olystyrene	- 0.28
Cu/MgO/KCl	- 0.17
Au/Si/KCl	- 0.81
Au/Polystyrene/H$_2$O	- 0.14

The table 1 above consists in systems that the individual Lifshtiz-Hamaker constants obey. This therefore demonstrates that the concept of negative van der

Waals force is physically sound. For a system of a particle of radius R, interacting with plane solid surfaces in a liquid medium, Eq.(9) can be written as;

$$\Delta F^{adh} = -\left[\frac{A_{PLS}R}{6\pi d_0^2}\right]$$

(10)

Because of the problem of establishing the radius R of the virus, in this preliminary study, it is assumed that on a molecular level, the two surfaces approaching themselves appear plane to each other, so that Eq.(9) can be used. The Hamaker coefficient is related to the interfacial free energies by;

$$A_{PLS} = -12\pi\, d_0^2\, (\gamma_{PS} - \gamma_{PL} - \gamma_{SL})$$

(11)

This is obtained by combining Eq.(4) with Eq.(9). It is worth noting that for interaction of a particle against itself, then from Eq.(8);

$$A_{ij} = \frac{3}{4}\pi\hbar \int_0^\infty \left[\frac{\varepsilon_i(i\zeta) - \varepsilon_j(i\zeta)}{\varepsilon_i(i\zeta) + \varepsilon_j(i\zeta)}\right]^2 d\zeta$$

(12)

Thus, for our system,

$$A_{PLP} = A_{PP} + A_{LL} - 2A_{PL} = (\sqrt{A_{PP}} - \sqrt{A_{LL}})^2$$

(13)

$$A_{PLS} = A_{PS} + A_{LL} - A_{PL} - A_{SL} = (\sqrt{A_{PP}} - \sqrt{A_{LL}})\,(\sqrt{A_{SS}} - \sqrt{A_{LL}})$$

(14)

To determine the combined Hamaker coefficient using the Lifshitz theorem of Eq.(12), there is a need to evaluate the dielectric constant ε of that equation. This could be done through the measurement of the absorbance for each sample of infected and uninfected blood. From the knowledge of light absorbance, reflection and transmittance, it could be noted that;

$$\bar{a} + T + R = 1$$

(15)

Where; $\bar{a}$ is absorbance, T is transmittance, and R is reflectance.

Also;
$$T = \exp^{-\bar{a}}$$

(16)

With the values of $\bar{a}$ and T ascertained, R could easily be derived by substituting into Eq.(15). The next step is to find a value for the refractive index, n employing the mathematical relation [14].

$$n = \left[\frac{1 - R^{\frac{1}{2}}}{1 + R^{\frac{1}{2}}}\right]$$

(17)

A value for the extinction coefficient, k is obtained from the equation;

$$k = \left[\frac{\alpha\lambda \times 10^{-9}}{4\pi} \right] \tag{18}$$

Where; α is the absorption coefficient defined as follows;

$$\alpha = \left[\frac{\bar{a}}{\lambda \times 10^{-9}} \right] \tag{19}$$

The dielectric constant, ε could thus be given by the formula [15].

For the real part; $\qquad\qquad\qquad \varepsilon_1 = n^2 - k^2 \tag{20}$

For the imaginary part; $\qquad\qquad\qquad \varepsilon_2 = 2nk \tag{21}$

With these values, it is possible to calculate A_{PLS} or A_{ij} using the relevant equations.

3. Research Methodology

3.1. Sample Collection

This research work involved collection of blood samples from twenty HIV infected and twenty uninfected persons. The blood samples were collected from Nnamdi Azikiwe University Teaching Hospital (NAUTH) Nnewi. Anticoagulant test tubes were used to ensure the freshness of the collected samples and to avoid the samples becoming lysed (spoilt).

3.2. Sample Preparation

The centrifugal separator was used to obtain such components as Red Blood Cells (RBC), White Blood Cells (WBC) also called the Lymphocytes or the Buffy Coat and the Plasma or Serum. The glass slides were prepared and smeared with the samples for absorbance measurements.

3.3. Measurements

The CD4+ counts of the blood samples were obtained using a digital CD4+ Counter. Absorbance measurements over a range of wavelength spanning between 230 and 890 Hertz were done with a digital Ultraviolet Visible Spectrophotometer (Ultrospec3100pro).

4. Data Analysis

4.1. *Plots from Collated Data*

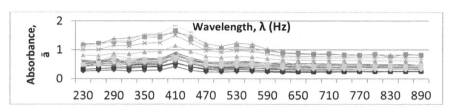

Figure 4. Plot of Absorbance, $\bar{a}$ versus Wavelength, λ for Twenty Samples of HIV Infected Red Blood Cells [16].

The plot in figure 4 reveals an interesting pattern for HIV positive red blood cells. The absorbance of the respective twenty HIV infected blood samples steeply increased as the wavelength increased until a critical wavelength of 410Hz, where a peak value was attained. A further increase in the wavelength saw at first a steep and latter a gradual decrease in the absorbance values. This peak value falls within the visible range of the ultraviolet radiation which is between 300 - 600Hz. The peak values of absorbance ranges between 0.424 and 1.832.

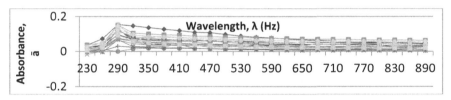

Figure 5. Plot of Absorbance, $\bar{a}$ versus Wavelength, λ for Twenty Samples of HIV Positive Lymphocytes (White Blood Cells) [16].

The plot in figure 5 shows a similar pattern as that of figure 4 with the peak value occurring at the wavelength of 290Hz. However, the peak absorbance values are of the range $0.019 \leq \bar{a} \leq 0.163$.

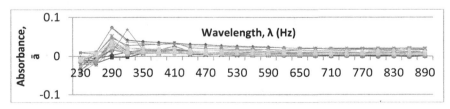

Figure 6. Plot of Absorbance, $\bar{a}$ versus Wavelength, λ for Twenty Samples of HIV Positive Plasma (Serum) [16].

Here once again the plot followed the initial pattern exhibited by other blood components. The peak value for this blood component occurred at a wavelength of 290Hz which corresponds exactly with that of the lymphocytes (WBC). However, the peak absorbance range is between 0.018 and 0.074. It is interesting though, that at the lower wavelengths of between 230–260Hz negative absorbance values were recorded.

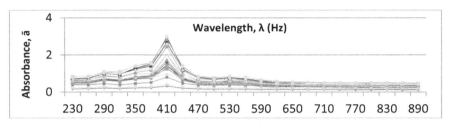

Figure 7. Plot of Absorbance, $\bar{a}$ versus Wavelength, λ for Twenty Samples of HIV Negative Red Blood Cells (RBC) [16].

The peak value of the absorbance for HIV negative red blood cells was obtained at the wavelength of 410Hz and ranges as follows $0.473\leq \bar{a} \gtrless 3$. The reaction here also follows the earlier patterns with all the various twenty samples showing moderately conformed characteristics.

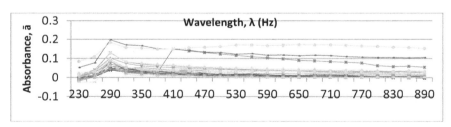

Figure 8. Plot of Absorbance, $\bar{a}$ versus Wavelength, λ for Twenty Samples of HIV Negative Lymphocytes or White Blood Cells (WBC) [16].

The plot for the samples of HIV negative lymphocytes reveals similar characteristics as their counterparts, however with the peak values occurring at 290Hz. The absorbance values at this peak point are of the magnitude between 0.040 – 0.197. The implication of these recorded peak values will later be explained.

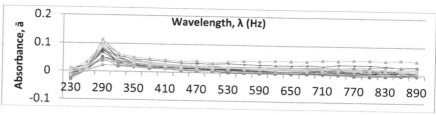

Figure 9. Plot of Absorbance, $\bar{a}$ versus Wavelength, λ for Twenty Samples of HIV Negative Plasma (Serum) [16].

For the HIV negative plasma samples the characteristics of the plot follow the usual pattern once more with the peak value obtained at a wavelength of 290Hz. The absorbance values at this peak point fall within the range of 0.021 and 0.114. This suggests that these wavelengths could be used as reference points for better results in subsequent researches and measurements (table 2).

4.2. Comparison between the Peak Absorbance Values of HIV Positive and Negative Blood Components

The decrease in the absorbance of the HIV infected blood samples reveals the role of the virus in significantly affecting the surface properties of the infected blood cells and specimens.

Table 2. Values of divers variables for both infected and uninfected blood components [16].

Blood Component	Wavelength at Peak Absorbance (Hz)	Ranges of Peak Absorbance	Cell Count ($\times 10^{10}$ Cells/l)	A_{ii} ($\times 10^{-21}$ J)
HIV Infected RBC	410	0.424 -1.832	230 - 450	2.9575
Uninfected RBC	410	0.474 - 3.000	470 - 560	3.1212
HIV Infected CD4 Lymphocyte	290	0.019 - 0.163	0.009 – 0.124	0.9868
Uninfected CD4 Lymphocyte	290	0.040 - 0.197	0.080 – 0.180	0.9659
HIV Infected Serum	290	0.018 - 0.074	-	0.2486
Uninfected Serum	290	0.021 - 0.114	-	0.4388

4.3. Computation of the Hamaker Coefficients

The values of the Hamaker constants obtained for the various blood components were employed in the derivation of the Hamaker coefficients, A_{132}. The underlying equation here is Eq.(22) with the necessary modifications as shown below.

$$A_{132} = \left(\sqrt{A_{11}} - \sqrt{A_{33}}\right)\left(\sqrt{A_{22}} - \sqrt{A_{33}}\right) \qquad (22)$$

The modifications are as follows;

$A_{11} = A_{11}$ values for HIV negative lymphocytes (WBC)

$A_{33} = A_{11}$ for HIV positive plasma (serum) as the intervening medium

$A_{22} = A_{11}$ values for HIV positive lymphocytes (WBC)

The infected lymphocytes are used in lieu of the virus because there is currently no known means of isolating the virus. The assumption here is that the infected lymphocyte is an approximation of the actual virus owing to the manner of the infection.

4.4. Deductions for the Absolute Combined Hamaker Coefficient A_{132abs}

Applying Lifshitz derivation for van der Waals forces as in Eq.(8), the absolute value for the Hamaker coefficient could be derived by obtaining mean of all the A_{132} values. An absolute combined Hamaker coefficient $A_{132abs} = 0.2587 \times 10^{-21}$ Joule was derived. To obtain a value for the combined Hamaker coefficient A_{131} for the uninfected blood the relation of Eqs.(23) and (24) are employed.

$$A_{131} = A_{11} + A_{33} - 2A_{13} \qquad (23)$$

$$A_{131} = \left(\sqrt{A_{11}} - \sqrt{A_{33}}\right)^2 \qquad (24)$$

Upon obtaining the mean of all values of A_{131} for the twenty uninfected blood samples, an absolute value A_{131abs} was derived as $A_{131abs} = 0.1026 \times 10^{-21}$ Joule. The near zero value of the A_{131abs} shows the absence of infection in the blood samples thus suggesting the usefulness of the concept of negative Hamaker coefficient in finding a solution to HIV infection.

5. Conclusion and Recommendation

5.1. Conclusion

- This novel research work on HIV-blood interaction has further buttressed the place of the relevance of engineering thermodynamics or at least quasi-thermodynamics in finding solution to various scientific and biological processes.
- The positive value of the absolute combined Hamaker coefficient $A_{132} = 0.2587 \times 10^{-21}$ Joules obtained for the HIV positive samples is an

affirmation that the blood samples were actually infected. This when compared with the absolute Hamaker coefficient $A_{131}=0.1026 \times 10^{-21}$ Joule obtained for the uninfected blood samples is conclusive of the fact of the relevance of the concept of Hamaker coefficient to HIV-blood interactions.

- This research concludes that there is a possibility of finding an antidote/cure for the HIV-AIDS pandemic if further work towards defining the conditions of the system that could render the absolute combined Hamaker coefficient negative and the additive(s) to the serum (in form of drugs etc.) as the intervening medium that could achieve this condition [16].

5.2. *Recommendation*

The following recommendations therefore are hereby made in furtherance of this research work.

- Further researches which will include the use of contact angle method as an alternative means of verification of the surface characteristics of blood be carried out.
- Efforts should be made towards the interpretation of the characteristics and specification of the material that would render the Hamaker Coefficient A_{132} negative as deduced in this research. This should involve a team of medical personnel like pharmacists, pharmacologists, laboratory scientists and doctors in collaboration with engineers and physicists. A synergy of this sort cannot be overlooked if a solution to the menace of HIV-AIDS could be in view soonest.

References

1. UNAIDS Report on the Global AIDS Epidemic/UNAIDS: *A UNAIDS 10th Anniversary Special Edition*, 1-2 (2006).
2. UNAIDS Intensifying HIV prevention: *Policy Position Paper: UNAIDS*, Geneva, (2005).
3. Omenyi, S.N., *The Concept of Negative Hamaker Coefficient: Nnamdi Azikiwe University, Awka, Inaugural Lecture Series*, **8.1**, 23 (2006).
4. Neumann, A.W., Absolom, D.R., Francis, D.W., et al, *Blood Cell and Protein Surface Tensions*, Annals New York Academy of Sciences, 277(1983).
5. Richard Hund, *Microbiology and Immunology On-line*, http://pathmicro.med.sc.edu/lecture/hiv7.htm , Retrieved 27-05-08.

314

6. Sheena McCormack, et al, *Science, Medicine, and the Future; Microbicides in HIV Prevention*, http://www.bmj.com/cgi/content/full/322/7283/410, Retrieved 24-05-08.

7. Sherman, I.A., Grayson, J. and Zingg, W., *The Critical closing Phenomenon in the hind limb of the dog,* Proc. 11th Europ. Conference Micro-Circulation, In Bibliotheca Anatomica, P. Gaehtgens Ed., S. Karger Basel, **20**, (1980).

8. Neumann, A.W, and Moy, E, *Colloids and Surfaces*, **43**, 349-365(1990).

9. Hough,D.H. and White, L.R., *Advanced Colloid Interface Science*, **14**, 3 (1980).

10. Hamaker, H.C., *Physica*, **4**, 1058 (1937).

11. Lifshitz, E.M., Dzyaloshinskii, I.E., et al, *Advanced Physics*, **10**, 165 (1961).

12. Israelachivili, J.N., *Proc. Royal Social Services* A, **331**, 39 (1972).

13. Visser, J., *Advances in Interface Science*, Elsevier Scientific Publishing Company, Amsterdam, **15**, 157–169 (1981).

14. Robinson, T.S., *Proc. Phys. Soc.* London, **65B910**, (1952).

15. Charles, Kittel, *Introduction to Solid State Physics*, 7th ed., John Willey and Sons Inc., New York, 308 (1996).

16. C.H. Achebe, S.N. Omenyi et al, *Human Immunodeficiency Virus (HIV)-Blood Interactions: Surface Thermodynamics Approach*, Lecture Notes in Engineering and Computer Science: Proceedings of The International Multi-Conference of Engineers and Computer Scientists 2012, IMECS 2012, 14-16 March, 2012, Hong Kong, 136-141.